SOCIAL NETWORK ANALYSIS

SOCIAL NETWORK ANALYSIS

with Applications

IAN A. McCULLOH
Curtin University

HELEN L. ARMSTRONG
Curtin University

ANTHONY N. JOHNSON
United States Military Academy

Published by John Wiley & Sons, Inc., Hoboken, New Jersey.
Published simultaneously in Canada.

For general information on our other products and services please contact our Customer Care
Department with the U.S. at 877-762-2974, outside the U.S. at 317-572-3993 or fax 317-572-4002.

Wiley also publishes its books in a variety of electronic formats. Some content that appears in print,
however, may not be available in electronic format.

Library of Congress Cataloging-in-Publication Data:

Social Network Analysis with Applications / McCulloh . . . [et al.].
   "Wiley-Interscience."
   Includes bibliographical references and index.
   ISBN 978-1-118-16947-6 (pbk.)
   1. Networks—Analysis.   2. Social
sciences—Research—Statistical methods. I. McCulloh II. Armstrong III. Johnson

HA31.2.S873 2007
001.4'33—dc22   2004044064

10 9 8 7 6 5 4 3 2 1

To Our Students

CONTENTS

PART I

NETWORK BASICS

PART II

SOCIAL THEORY

PART IV

ORGANIZATIONAL RISK

CHAPTER 9   *ORGANIZATIONAL RISK*   205

LIST OF FIGURES

LIST OF TABLES

FOREWORD

Readers of this book probably already realize that both popular attention to social networks and the systematic study of networks have exploded in the past decade or so. What might be less apparent is the historical depth and substantive breadth of the network perspective. Tracking down the very first scholarly use of the word "network" is much like locating the first living creature on earth—it had to be there somewhere, but once life (or the study of networks, for that matter) emerged, expansion and diversity took over and, in turn, provided the impetus for increasing diversity. The study of networks has followed a similar trajectory of diversification. As a maturing arena of inquiry, the field of networks has expanded from a somewhat arcane branch of mathematics (graph theory) and a relatively focused structural approach in the social sciences (sociometry and its descendants) to a powerful perspective for studying relational systems quite generally. Applications currently extend into essentially all domains of investigation: social structure and process, biological systems, formal organizations, computer networks, physical connections, semantic structures, genetic relations, neuronal networks, epidemiology of disease spread, diffusion of rumors, scientific knowledge networks, industrial collaborations, and international relations, to name a few. Despite this substantive diversity, researchers adopting a network perspective share a common conviction regarding the important consequences of interdependencies among units within their particular domains of interest. But beyond the focus on relations, the sheer omnivorousness of the network perspective means that it can, and does, admit a wide variety of applications.

Ian A. McCulloh, Helen L. Armstrong, and Anthony N. Johnson's book "Social Network Analysis with Applications" exemplifies the extraordinary potential of the network perspective with its specific focus on application of networks to organizational systems. In a crowded field of recent network books, McCulloh, Armstrong, and Johnson's stands out in providing an accessible introduction to network analysis, including substantive examples and interpretations grounded in organizational theories. Its practical applications to empirical data along with step-by-step instructions for analyzing networks using the Organizational Analyzer (ORA) software make it especially appropriate for newcomers to the study of networks.

One of the hallmarks of a maturing field is the translation of its core concepts and techniques into language that is readily accessible to newcomers and to practitioners who are primarily concerned with actionable results. McCulloh, Armstrong, and Johnson succeed admirably in that regard by drawing on their extensive combined experience teaching networks to soldiers, police officers, industry professionals, and both undergraduate and graduate students. Yet, Social Network

Analysis with Applications insists on maintaining the formal rigor required for serious network analysis and does not shy away from math and equations when necessary to convey a concept. The book's unifying theoretical perspective and interpretative examples center on organizational risk and intervention. This focus draws attention to the consequences of particular organizational forms and individual positions and leads to consideration of how network actions and interventions might affect desirable or undesirable outcomes. This focus will surely provoke readers to consider the vast potential for applying network analysis in real-world settings.

I invite you to explore this book, to learn from it, and to use its many insightful examples and analytical approaches to expand your network imagination!

Department of Sociology and Institute   KATHERINE FAUST
for Mathematical Behavioral Sciences
University of California, Irvine

PREFACE

We are very pleased to present "Social Network Analysis with Applications." This book will focus on models and methods for social network analysis applied to organizational risk. Current books in the area of social network analysis are both highly technical and written at the advanced graduate level or they only discuss general concepts and omit mathematical calculations. Very few of the texts offer any practice problems for practitioners and students to complete as practice problems or homework exercises.

The inclusion of mathematical calculations is central to our approach in this text. Many in the field prefer to present social network analysis by hiding the mathematics and relying on computer software to identify centrality. We contend that this is a critical mistake in the pedagogy of network analysis. The authors have taught over 30 courses in social network analysis to over 500 students with varied approaches and consulted with many more colleagues. Those who have learned how to calculate centrality measures by hand calculation, for example, are 11 times more likely to retain an understanding of what the measures mean 1–3 months after the course. Thus, it is not our contention that an individual would use hand calculations on any real-world example. However, in learning to calculate measures by hand, the mathematics leads to an understanding of the underlying principles of social network analysis.

Many practitioners in industry, management, military intelligence, and law enforcement have expressed a growing interest in social network analysis, specifically focused on identifying organizational risk. We operationally define organizational risk as vulnerability in the social network. This could be a node high in informal power or a rare broker of resources. This could be a point of influence for the diffusion of ideology. There may exist many networks within an organization, such as a friendship network, a resource network, or a knowledge network. One or more of these networks may present organizational risk, while the others do not. In a military or law enforcement application, organizational risk identifies targets for further development and investigation. In an industry or management application, organizational risk identifies informal power brokers that should be included in management decisions, and potential vulnerability from lack of redundancy. The authors attempt to present examples of both.

The authors have trained soldiers in Iraq and Afghanistan, local and national police, and industry professionals on the applications of social network analysis. The topics laid out in this book follow the curriculum that they have used for the past 7 years in teaching week-long workshops as well as graduate and undergraduate level college courses.

The first three chapters introduce the mathematical concepts of a network and centrality. Again, we contend that an understanding of the mathematics leads to a deeper and more complete understanding of the social concepts behind social network analysis. Social network analysis software is also presented for larger problems. Visual analysis methods are also introduced. At the conclusion of the first three chapters, the reader should have a basic understanding of common network analysis techniques.

Chapters 5–7 provide the social theory that underlies social network analysis. It is infeasible to include all social theory related to social networks in the space afforded in this text. Therefore, the social theory that is most applicable to understanding organizational risk is provided. The authors' selection of the material included in these sections comes from experience applying social network analysis in industry, counter terrorism, and law enforcement. We have also sought input from former students who are actively using social network analysis on a regular basis. The reader is reminded that the focus of this text is for practitioners intending to apply social network analysis to organizations.

Chapters 7 and 8 are directed toward data. Matrix algebra is included in an appendix and provides a primer for the necessary mathematics to understand meta-networks and relational algebra. Relational algebra is an often overlooked method. While there is limited application for single mode networks, it is a critical tool for handling meta-networks, also known as *multiplex* or *multimode networks*. Relational algebra is the means to transform any relational data into different social networks, each of which might reveal organizational risk. Chapter 8 reviews sources of data. Strengths and weaknesses of data sources are presented. At the conclusion of Chapter 8, the reader should be capable of collecting network data and identifying organizational risk. The final chapter provides an organized approach to applying the methods and techniques presented in the book. It is intended to serve as a review, leading the reader through examples of applied social network analysis.

We are confident that you will enjoy "Social Network Analysis with Applications," and feel empowered by the time you are finished. We have been amazed at the innovative and interesting ways our students have applied social network analysis to a wide variety of problems. More are sure to follow. Welcome to the fastest growing field in science!

IAN A. MCCULLOH
Curtin University, Australia

September, 2013

ACKNOWLEDGMENTS

There are many people who have contributed to our understanding of social networks and the culmination of this text. For those interested, we present our journey of learning, while acknowledging those who made it possible.

In 2005, while conducting research in response surface optimization for the Army, I heard John Parmentola state, "Network Science is the Army's number four research priority." As I began to learn more about this exciting field, I discovered several presentations from Steve Borgatti posted online. Steve is an amazing teacher and has the ability to present the most complex material in a very simple way. He was very open and collaborative with me and allowed me to use his posted materials when I taught my first social network analysis course at West Point in 2005.

Two students emerged out of my initial entry into social network analysis, Julie Paynter and Victor Basher. Had it not been for these two students, I probably would not have continued into the discipline. Julie was ranked number three in the class of 2006 at West Point. She was going to be a military intelligence officer and was majoring in mathematics. She had done an internship the summer before studying influence in jihadist texts. Her senior thesis applied social network analysis to understand influence in jihadist authors in Iraq. Vic Basher had been an intelligence analyst with the 10th Special Forces Group (Airborne) in Iraq and subsequently went to West Point. He was in my basic probability and statistics course, having difficulty understanding why an Army officer needed to learn math and physics. He was independently downloading insurgent propaganda videos on his personal computer to look for forensic clues that could help his friends who were at the time deployed to Iraq. In an effort to motivate his study of mathematics, I suggested an application of social network analysis to his data. As of April 1, 2006, Vic had downloaded 74% of all US deaths in Iraq and Afghanistan and had cross-referenced them with hometown news reports. The network we made allowed us to identify clusters (using one of Steve Borgatti's posted lectures) that helped us identify insurgent groups. Vic, Julie, and I presented our research at a variety of military audiences to include the Defense Intelligence Agency (DIA), National Security Agency (NSA), the newly formed Joint Improvised Explosive Device Defeat Organization (JIEDDO), the Asymmetric Warfare Group (AWG), and Special Operations Command (SOCOM).

During my tour of military agencies, I met Kathleen Carley. She told me that she was interested in collaborating with me. My brash response was that I "needed to get a Ph.D. out of the deal." She put up with my arrogance. She taught me more than any professor or teacher I have ever known. She is one of the most brilliant academics I have ever met. Her contributions to the field of social network analysis are so advanced; many have difficulty believing they aren't science fiction.

Few will ever know how many lives she saved through her contributions to the US Military's Global War on Terror. She showed the military how to target threats correctly, collect evidence, and prevent innocents from being erroneously targeted. I am very proud that she was willing to serve as my advisor.

Around the same time, I received my first large research grant in the area of social networks. Joe Psotka and Dan Horn at the US Army Research Institute for the Behavioral and Social Sciences awarded me a grant to study and compare email networks with face-to-face friendship networks. In addition, they provided seed money to establish West Point's Network Science Center. As we stood up this center, my good friends Tony Johnson and Helen Armstrong took our efforts to a new level, expanding our research directions, organizing conferences, winning grants, and most importantly, teaching. Through our early conferences we met Nosh Contractor, László Barabási, and Guido Calderelli. Each of these individuals showed us different and novel applications of network analysis in social media, biology, physics, and fractals. They showed us how to reach out across our academic institution and create truly interdisciplinary and collaborative research.

Another student challenged my world-view of academia, Josh Lospinoso. I was the Mathematics Department academic advisor. Josh had recently majored in Operations Research and Tony and I couldn't understand why he had no interest in physics or engineering. He told us, "Those were easy problems that anyone could solve." He was interested in the much more challenging and complex issues of human and social behavior; but not from a qualitative perspective. He wanted to model it mathematically. Josh joined Tony, Helen, and I as we learned about social networks together. Josh interned every summer at the NSA, applying social networks to defense applications as a cadet. Some argue that the success in Iraq in 2006–2007 had less to do with the surge of forces and more to do with the successful targeting of threats. Josh's work contributed to that effort.

We learned that Josh was not alone in his view that social science provided a challenging application area for mathematics. We discovered mathematical sociology and anthropology. Through the Sunbelt conferences we were able to meet Russ Bernard, Dimitrios Christopolous, Jeff Johnson, David Krackhardt, and many others who encouraged us in our research and motivated us to bring social network analysis to the undergraduate student populations we had at West Point. Tom Valente, in particular, had many conversations with us about teaching social network analysis. He not only shared his materials freely but helped us shape the way we present material and the content of our courses. He has even organized sessions at Sunbelt for professors to discuss the pedagogy of social network analysis.

Through the support and collaboration of all of these individuals, our program at West Point grew from Julie and Vic to seven undergraduate students presenting research at the 2007 Sunbelt conference in Corfu, Greece, to 23 students at the 2011 Sunbelt conference in Tampa, Florida. There are now well over a dozen faculty and 100 undergraduate students studying social network analysis at West Point. Helen established the Centre for Organizational Analysis in the School of Information Systems at Curtin University of Technology in Perth, Australia, which expands this effort and it continues to grow. Ian and Tony have worked to expand an understanding of social network analysis throughout the US Military to include teaching 16 courses in Iraq and Afghanistan. We are grateful to the International Network of Social Network Analysts (INSNA) as an academic community, for being so receptive to us and enabling us to develop this text and encourage us in our academic endeavors.

I. A. M.

INTRODUCTION

The analysis of networks is not a recent trend. With historical examples dating back several centuries, its use in scientific enquiry has increased over the past few decades in particular. The term *network* can refer to any number of different types of networks, for example, social networks of people or a network of roads or a computer network. The Oxford dictionary defines a network as a group or system of interconnected people or things, suggesting that a network contains objects or people, and connections or links between these. By studying the objects, the links, and also the structure and dynamics of the network we can discover many important aspects that were previously not known. In this text, we focus on social network analysis. Linton Freeman describes the discipline.

"The social network approach is grounded in the intuitive notion that the patterning of social ties in which actors are embedded has important consequences for those actors. Network analysts, then, seek to uncover various kinds of patterns. And they try to determine the conditions under which those patterns arise and to discover their consequences" (Freeman, 2004).

People do not act in a manner independent from the context of their social interactions and environment. Moreover, there are patterns and behaviors of interaction that are common among people, if one can properly understand the social context in which they occur. This is not restricted to social ties between people. It may include knowledge, resources, tasks, beliefs, roles, organizations, and more, as well as peoples' interactions with these other factors. To better understand what social network analysis is, we provide Freeman's definition of four features that make up social network analysis:

1. Social network analysis is motivated by a structural intuition based on ties linking social actors.

2. It is grounded in systematic empirical data.

3. It draws heavily on graphic imagery.

4. It relies on the use of mathematical and/or computational models.

"Beyond commitment to these four features, however, modern social network analysts also recognize that a wide range of empirical phenomena can be explored in terms of their structural patterning" (Freeman, 2004).

This structural perspective in the social sciences dates back to the establishment of the field. Comte (Martineau, 1895) identified two branches of sociology, statics, and dynamics. He further defined statics as the study of the "laws of social interconnection." This is essentially the theoretical roots of social network analysis. The appeal of the network approach to problems has also appealed to others in a broad range of disciplines from physics to the other social sciences. As such, there are methods and tools that can be exchanged and applied to a wide variety of problems in many different fields. This has greatly enriched the development of the science and the contributions social network analysis has made to many different fields. Others in the early development of social science introduced data, methods, and theory, which we now recognize as social network analysis. Simmel (1909) defined society as existing "where a number of individuals enter into interaction." The focus of the research of Simmel and his students was on the structure of social patterns.

Hobson, a famous economist at the turn of the twentieth century, presented an affiliation matrix on overlapping directorships of financiers in South Africa. There were five companies and six individuals. This was an early attempt at relational algebra, which we present in Chapter 7. This represents an important step forward in the ability to recognize and construct social network data. It also highlights how non-people nodes can impact our representation of a social network.

We find it interesting that the first evidence of social network data was derived from relational data rather than direct observation, which was not apparent in the literature until the 1920s. Systematic observation of social network data was introduced by Wellman (1926) who recorded data of preschool children during play. Blatz and Bott (1928) identified different forms of interaction among children. This enabled her to focus her observations on interaction specific to a research context.

The idea of different types of interaction, or different types of links, was the beginning of the formation of the underlying theory of meta-networks, proposed by Kathleen Carley years later. Meta-networks are discussed in Chapter 7. Recently, people have begun to refer to a subset of meta-networks as multilevel networks. In addition, Bott would focus on one child at a time to observe the different types of interaction with others. She recorded these data in matrices. Hagman compared the data collection approaches of observation and interview to reveal differences and bias in findings (Hagman, 1933). This became an important issue in social network analysis, studied extensively by Bernard and Killworth in a series of excellent papers on the topic. This issue is discussed in Chapter 8.

Most people credit the beginnings of social network analysis to Jacob Moreno with the graphical social network of interactions between school children in the New York Times (Moreno, 1932, 1934) displayed in Figure I.1. Moreno initially referred to his research as "psychological geography," but later changed the name to sociometry and started a journal of the same name in 1938. Moreno's work was largely theoretical and his social networks were more illustrative than mathematical representations. Recognizing this, Moreno teamed up with one of the earliest mathematical sociologists from Columbia University, Paul Lazarsfeld. Lazarsfeld developed a probabilistic model of a social network, which appeared in the first volume of Moreno's journal.

Other groups began to explore social network analysis and structural approaches to social behavior. Lewin was a well-known sociologist who embraced the structural approach. He developed the Research Center for Group Dynamics at MIT with four of his former students including Cartwright and Festinger. Following Lewin's sudden death and MIT's decision to close the research

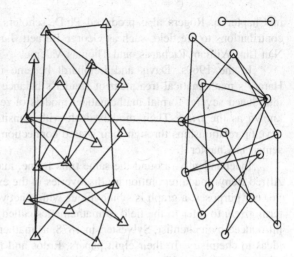

EMOTIONS MAPPED BY NEW GEOGRAPHY

**Charts Seek to Portray the
Psychological Currents of
Human Relationships.**

New York Times
April 3, 1933

FIGURE I.1   Emotions mapped by new geography.

center, the Center moved to the University of Michigan in 1948. Cartwright and Festinger recruited Frank Harary, who had earned his Ph.D. and was teaching in the Department of Mathematics, as a mathematical collaborator. Cartwright and Harary proposed a mathematical statement of the notion of cognitive balance proposed by Heider, which we discuss in Chapter 4. Others working in this research center conducted work on personal influence and the diffusion of rumors. In the 1950s, this group collected a famous multiyear data set on affinity in a university dormitory (Newcomb, 1961).

The Columbia University also made several important contributions to social network analysis. Robert Merton was a sociologist with a background in theory who teamed up with Paul Lazarsfeld, who was a mathematician working as a methodologist. They worked together throughout their careers, mainly in the areas of communication and what developed into market research. They also produced many Ph.D. students who became very influential in the field. Menzel and Katz (1956) investigated the diffusion of drug information among physicians, shaping our understanding of marketing. Peter Blau (1977) proposed homophily, which states that people with similar characteristics are more likely to meet. We present homophily in Chapter 4. Charles Kadushin developed the concept of social circles, which paved the way for subgroup analysis, which we discuss in Chapter 5.

Everett Rogers is another important figure in social network analysis during this time. His dissertation from Iowa State University involved some simple social network analysis methods in the study of the diffusion of innovations. He later introduced the concepts of early adopters and the bell curve and S-curves for the adoption of innovations. He extended his research to preventative health and other applications. As of the writing of this book, Rogers' book on "The Diffusion of Innovations" is the second most cited book in the social sciences. Noah Friedkin and Kathleen Carley extended Rogers' work to the diffusion of ideology, which is important for understanding radicalization. These concepts are presented in Chapter 6. Rogers also produced Ph.D. scholars who have made impressive contributions to the field, such as George Barnett, James Danowski, Peter Monge, Nan Lin, William Richards, and Thomas Valente.

In the 1960s, Davis and Leinhardt became interested in Cartwright and Harary's mathematical treatment of Heider's balance theory. Davis and Leinhardt introduced several formal mathematical models of relations between three actors, known as the *triad*. Their models dealt with transitivity, which is where actors develop relationships through their shared connection to others. Their work is presented in Chapter 4.

Independently, around the same time frame, mathematicians Paul Erdős and Alfréd Rényi made revolutionary discoveries in the evolution of "random graphs." For our purposes, a graph is synonymous with a network. Erdős and Rényi use the term *graph* to refer to the field of mathematics called *graph theory*. This term was introduced by a chemist, Sylvester, in 1878, as mathematicians were applying their ideas to chemistry. In their eight papers, Erdős and Rényi evaluate the properties of random graphs with n nodes and l links. For a random graph, G, containing no links, at each time step a randomly chosen link among the possible links is added to G. All of the possible links are equiprobable. A general model used to generate random graphs is as follows: "For a given p, $0 \leq p \leq 1$, each potential link of G is chosen with probability p, independent of other links. Such a random graph is denoted by $G(n,p)$, where each link is determined by flipping a coin, which has probability p of coming up heads." In this model of random graphs, each link has an equal probability of occurring or not occurring within the graph. This random graph model also assumes that all nodes in the graph are present at the beginning and the number of nodes in the network is fixed and remains the same throughout the network's life. In addition, all nodes in this model are considered equal and are undistinguishable from each other.

Building upon Cartwright and Harary's work, and utilizing Erdős' theory of random graphs as well as the class of uniform distributions associated with these graphs, Holland and Leinhardt (1981) developed a variety of statistical tests for the analysis of social networks. Using a uniform distribution, these tests spread the total probability mass equally over all possible outcomes, thereby giving an equal probability to the existence of a link between any two nodes in the network. These statistical tests were used to develop a reference frame or constant benchmark to which observed data could be compared in order to determine how "structured a particular network was, or how far the network deviated from the benchmark" (Wasserman and Faust, 1994). While this text does not address the statistical treatment of networks, it is important for interested readers to be aware that there are well-developed statistical tests for networks.

One of the most important figures in social network analysis was Harrison White. Harrison was a physicist who became interested in social behavior. He was working at Carnegie Tech, where he met Nobel laureate Herbert Simon. Herbert Simon was a founding father of many scientific fields to include artificial intelligence, organization theory, complex systems, and computer simulation. He was also the first to propose preferential attachment as a mechanism to explain power law distributions in the 1950s. Simon's influence on Harrison White led him to pursue a second Ph.D. in Sociology from Princeton. Harrison then moved to Harvard, where he produced some of the most important leaders in the field of social network analysis including Peter Bearman, Paul Bernard, Phillip Bonacich, Scott Boorman, Ronald Breiger, Kathleen Carley, Mark Granovetter, Joel Levine, and Barry Wellman. He introduced blockmodeling to understand subgroups and interactions between them, which we explore in Chapter 5. He proposed the concept of structural equivalence. These concepts are foundational to understanding network structure. Linton Freeman states, "Contemporary network analysis could never have emerged without Harrison White's contributions" (Freeman, 2004).

Linton Freeman is another prolific scholar in the field of social network analysis. He defined the network centrality measures that form the core metrics of social networks. These measures provide mathematical expressions of informal power, diffusion reach, and direct influence in a network. These concepts represent the foundation of modern social network analysis and are presented in Chapters 2 and 3. Freeman also produced many Ph.D.s who have made numerous important contributions to the field, including Stephen P. Borgatti, Katherine Faust, Sue C. Freeman, Jeffery Johnson, David Krackhardt, and Lee Sailer.

Equally important to the development of social network analysis was the growth of a community of scholars interested in the field. Their collaboration was facilitated greatly by H. Russell Bernard and Alvin Wolfe with their organization of the Sunbelt conferences. Initially, Russ received a grant while he was teaching at the West Virginia University to host a small meeting of mathematicians, sociologists, and anthropologists working in the area of social network analysis in 1974. Several follow-on meetings occurred. By 1980, Russ had moved to the University of Florida and developed a friendship with Alvin Wolfe at the University of South Florida who was also interested in social network analysis. They decided that Florida was a nice place and that people would want to visit, especially to discuss networks. So they organized the first Sunbelt Conferences in Tampa, Florida, in February of 1981 and 1982. They established three rules. There would be no sessions during peak tanning hours. The older guys would sponsor a hospitality social for the younger guys to facilitate interaction in an informal setting. The conferences would be held in a warm, enjoyable place during the winter to attract people to attend. The Sunbelt conferences have grown in size significantly over the years and they now hold sessions throughout the day in multiple parallel tracks; however, it remains one of the most collaborative and accessible conferences in academia, bringing scholars across many disciplines together to collaborate on social network analysis. Later, Barry Wellman formed the academic professional society known as the "*International Network of Social Network Analysts*" (*INSNA*), which now organizes Sunbelt as its official annual meeting as well as editing the journal *Connections*. The final major development in social network analysis was the development of computer software to enable the mathematical calculations of network structure. Linton Freeman teamed up with Martin Everett and Steve Borgatti to produce UCI Net, which is one of the most widely used social network analysis software programs. Pajek is a European software tool that is well suited for displaying large data sets. Kathleen Carley developed the Organizational Risk Analyzer (ORA), which is designed to analyze meta-networks with multiple types of nodes and links, as well as a variety of visualization and overtime analysis features. The practical labs in this text are demonstrated in ORA. Recently, an open-source statistics package, **R**, featured powerful network analysis tools to include exponential random graph modeling with the *statnet* package and actor-oriented modeling with the *Rsiena* package. There are of course many other social network analysis tools. Social network analysis software has certainly enhanced the capability and practical application of the field.

Social network analysis offers more than pictures. It provides an entirely new dimension of analysis for organizational behavior. Traditional analysis focuses on individual attributes. Social networks focus on relationships between individuals. Traditional analysis assumes statistical independence, where social network analysis focuses on dependent observations. Traditional analysis seeks to identify the correlation between significant factors and a response variable. Social network analysis seeks to identify organizational structure. The underlying mathematics behind traditional analysis is calculus, the language of change. The corresponding mathematics behind social network analysis is linear algebra and graph theory. These differences can be significant in terms of how someone looks at social dynamics. We hope you are empowered by this treatment of an exciting and powerful approach to social science.

REFERENCES

Blatz, W. E. and Bott, H. M. (1928). *Parents and the Pre-School Child*. J. M. Dent.

Blau, P. M. (1977). *Inequality and Heterogeneity: A Primitive Theory of Social Structure*. Free Press.

Freeman, L. (2004). *The Development of Social Network Analysis: A Study in the Sociology of Science*. Empirical Press.

Hagman, E. (1933). *The Companionships of Preschool Children, by Elizabeth Pleger Hagman, ... [Foreword by George D. Stoddard.]*. University.

Holland, P. and Leinhardt, S. (1981). An exponential family of probability distributions for directed graphs. *Journal of the American Statistical Association*, **76**(373):33–50.

Martineau, H. (1895). *La Philosophie positive d'Auguste Comte*. Luis Bahl.

Moreno, J. L. (1932). *Application of the Group Method to Classification*. National Committee on Prisons and Prison Labor.

Moreno, J. L. (1934). *Who Shall Survive?* Nervous and Mental Disease Publishing Company.

Newcomb, T. M. (1961). *The Acquaintance Process*. Holt, Rinehart and Winston.

Simmel, G. (1909). The problem of sociology. *The American Journal of Sociology*, **15**(3):289–320.

Wasserman, S. and Faust, K. (1994). *Social Network Analysis: Methods and Applications*. Cambridge University Press.

Wellman, B. (1926). *The Development of Motor Co-ordination in Young Children, an Experimental Study in the Control of Hand and Arm Movements, by Beth Wellman, ...* the University.

PART I

NETWORK BASICS

CHAPTER 1

WHAT IS A NETWORK?

> Injustice anywhere is a threat to justice everywhere. We are caught in an inescapable network of mutuality tied in a single garment of destiny. Whatever affects one directly affects all indirectly.
>
> —Martin Luther King Jr., Letter from a Birmingham jail, April 16, 1963

Learning Objectives

1. Understand social network terminology.
2. Construct an adjacency matrix.
3. Construct a network from given data.

In the movie, Good Will Hunting, the main character, Will Hunting, played by Matt Damon, is a janitor at a university. As he is mopping floors, he notices a problem posted on a board as a challenge to math students. The solution to the first two parts of the problem uses network analysis.

First Two Parts of the Good Will Hunting Problem

Let G be Figure 1.1 on the right.

- Find the adjacency matrix **A** of the graph G.
- Find the matrix giving the number of three-step walks for the graph G.

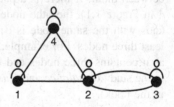

FIGURE 1.1   Graph G.

Social Network Analysis: with Applications, First Edition.
Ian A. McCulloh, Helen L. Armstrong, and Anthony N. Johnson.
© 2013 John Wiley & Sons, Inc. Published 2013 by John Wiley & Sons, Inc.

1.1   BASIC NETWORK CONCEPTS

A network is a collection of points linked through some type of association. These points can represent any object or subject (e.g., people, places, and resources) and the links can represent any relationship between them (e.g., route, distance, family membership, and reporting structure). The network is graphically illustrated using lines, arcs, and symbols so the viewer can visualize and analyze the structure of the network more easily. A simple network of four points can be seen in Figure 1.1. In this network the points are people and the links are relationships between the people.

A *graph* is the visual representation of a set of points, frequently called *vertices* or *nodes* that are connected by line segments called *edges* or *links*. Social *networks* are graphs that contain a finite set or sets of actors which we call *agents* and the relation or relations defined between them. A social network would then be comprised of nodes representing people with the corresponding links representing the relationship between the people.

It is important to understand how to navigate through a network graph. The information gained can help us understand how information flow through a network can be used to analyze the placement of nodes within the network and gauge their significance. We will look at how to do this analysis in the next few chapters, but first we need to understand network graph navigation terminology. Figure 1.1 has four vertices or nodes. Each node is a person. Moving from one node to another along a single edge or link that joins them is a *step*. A *walk* is a series of steps from one node to another. The number of steps is the *length* of the walk. For instance, there is a walk of three steps from node 1 to node 3 using the steps 1 to 4, 4 to 2, and 2 to 3. A *trail* is a walk in which all the links are distinct, although some nodes may be included more than once. The length of a trail is the number of links it contains. For example, the length of the trail between nodes 3 and 4 is 2, where 3 to 2 is the first link, and 2 to 4 is the second link. A *path* is a walk in which all nodes and links are distinct. Note that every path is a trail and every trail is a walk. In application to social networks, we often focus on paths rather than trails or walks. An important property of a pair of nodes is whether or not there is a path between them. If there is a path between nodes n_i and n_j (say nodes 1 and node 4 in Figure 1.1), then the nodes are said to be *reachable*. A walk that begins and ends with the same node is called a *closed walk*. A *cycle* is a closed walk of at least three nodes. For example, the closed walk 1 to 4, 4 to 2, and 2 to 1 is a cycle as it contains three nodes and begins and ends with node 1. Cycles are important in the study of *balance* and *clusterability* in signed graphs (a topic we explore later in the book).

1.2   ADJACENCY MATRICES, GRAPHS, AND NOTATION

We have seen that social networks can be graphically depicted as graphs. Social networks can also be represented mathematically by matrices. For example, Figure 1.2 is a *friendship graph* representing friendship between a group of people. We can

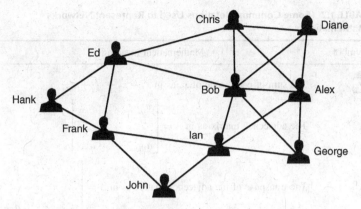

FIGURE 1.2   A social network of 10 people.

TABLE 1.1   Adjacency Matrix of a Friendship Graph

A	Alex	Bob	Chris	Diane	Ed	Frank	George	Hank	Ian	John
Alex	0	1	1	1	0	0	1	0	1	0
Bob	1	0	1	1	1	0	1	0	1	0
Chris	1	1	0	1	1	0	0	0	0	0
Diane	1	1	1	0	0	0	0	0	0	0
Ed	0	1	1	0	0	1	0	1	0	0
Frank	0	0	0	0	1	0	0	1	1	1
George	1	1	0	0	0	0	0	0	1	0
Hank	0	0	0	0	1	1	0	0	0	0
Ian	1	1	0	0	0	1	1	0	0	1
John	0	0	0	0	0	1	0	0	1	0

[a] Adjacency matrices need not always contain binary data. They can also be weighted conveying the strength of a particular tie.

record how the people in the graph are related in a mathematical object known as an *adjacency matrix*. The adjacency matrix in this case is a square (meaning same number of rows as columns) agent by agent matrix. Table 1.1 is the adjacency matrix for Figure 1.2. Mathematically, any network can be represented by an adjacency matrix, denoted by **A**, which, in the simplest case, is a $n \times n$ symmetric matrix, where n is the number of vertices or nodes in the network. The adjacency matrix is comprised of elements

$$a_{ij} = \begin{cases} 1 & \text{if an edge exists between } n_i \text{ and } n_j \\ 0 & \text{if no edge is present between } n_i \text{ and } n_j \end{cases}$$

We often represent the entire matrix with a capital letter **A**, **B**, or **C**. For the purpose of our friendship example, we will call the adjacency matrix **A**. The contents of any given cell are denoted a_{ij}. For example, in the matrix above, $a_{12} = 1$, because Alex is linked to Bob. This matrix is *symmetric* but it does not have to be (a_{ij} is

TABLE 1.2 Some Common Notations Used to Represent Networks Mathematically

Symbol	Mathematical meaning
$\sum\limits_{i=1}^{n} x_i$	The sum of each value of x, as in $x_1 + x_2 + \cdots + x_n$
$\mathbf{A}$	The adjacency matrix as in $A = \begin{bmatrix} a_{ij} & \cdots & a_{iN} \\ \vdots & \ddots & \vdots \\ a_{Nj} & \cdots & a_{NN} \end{bmatrix}$
$\mathbf{A}^{\mathrm{T}}$	The transpose of the adjacency matrix as in $A = \begin{bmatrix} a_{ij} & \cdots & a_{Nj} \\ \vdots & \ddots & \vdots \\ a_{iN} & \cdots & a_{NN} \end{bmatrix}$

$^a \mathbf{A}^{\mathrm{T}}$ is sometimes denoted $\mathbf{A}'$.

not always equal to a_{ji}). When the matrix is symmetric, it means that relationships are reciprocated. This is called an *undirected network*. Symmetric matrices can be used to represent friendship, distance, conversations, and similarity in attitude, among other relationships. *Directed networks* are not symmetric and can be used to represent friendship that is not necessarily reciprocated, transfers of resources, and authority (such as a chain of command), among other relationships. Anything that can be represented as a graph, can also be represented as a matrix.

By convention, data is represented so that the row agent is related "to" the column agent. For example, if the relation is "trust," then $a_{ij} = 1$ means that agent i trusts agent j, but if $a_{ji} = 0$ then agent j does not trust agent i in return. The transpose of a matrix $\mathbf{A}$ is denoted $\mathbf{A}^{\mathrm{T}}$. The transpose simply interchanges the rows with the columns. This will be discussed in greater detail later in the book. Table 1.2 is a table of some of the notation used to describe networks.

1.3 NODES AND LINKS

So far we have only used nodes to represent people; however, almost anything can be represented as a node or vertex. Nodes can be cities, equipment, organizations, resources, knowledge, tasks, beliefs, or any other object. A node can be anything we define it to be. Nodes that are all of the same thing (i.e. cities, equipment, organizations, etc.) are said to be of the same *node class*. Nodes can also exhibit attributes: a person node can have attributes such as age, gender, and the like; cities may have attributes such as population and location.

For a network to exist, the nodes must be linked by some kind of flow or relationship. Again, the link may be defined as anything that meaningfully represents the relationship. Social relations can be thought of as dyadic attributes. That is, attributes that characterize two nodes. While mainstream social science is concerned with monadic (one node) attributes (e.g., income, age, and sex), network analysis is concerned with attributes of pairs of individuals, of which binary relations are the main kind. Some examples of dyadic attributes are:

- Kinship: brother of, father of
- Social roles: boss of, teacher of, friend of
- Affective: likes, respects, hates
- Cognitive: knows, views as similar
- Actions: talks to, has lunch with, attacks
- Flows: number of cars moving between
- Distance: number of miles between
- Co-occurrence: is in the same club as, has the same color hair as
- Mathematical: are two links removed from

The relationship between two nodes may be either *undirected* or *directed*. Table 1.3 and Table 1.4 are the adjacency matrices for Figures 1.3a and 1.3b, respectively. An undirected relationship, Figure 1.3a, means the link is reciprocal, such as "married to," "trades with," or "miles between." A directed relationship,

TABLE 1.3   UnDirected Matrix

	A	B	C	D	E
A	0	1	1	0	0
B	1	0	1	0	0
C	1	1	0	1	1
D	0	0	1	0	0
E	0	0	1	0	0

TABLE 1.4   Directed Matrix

	A	B	C	D	E
A	0	1	1	0	0
B	0	0	0	0	0
C	0	1	0	1	0
D	0	0	0	0	0
E	0	0	1	0	0

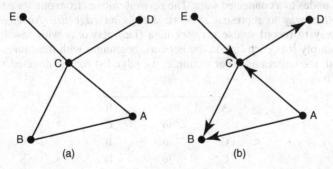

FIGURE 1.3   (a) Undirected and (b) directed graphs.

Figure 1.3b, is where there is a movement in only one direction and this direction is indicated with an arrowhead. Examples of directed relationships are "respects," "boss of," and "seeks advice from." Not all relationships are two way. For example, in Figure 1.3b, node A may seek advice from node C in a network, but node C may not seek advice from node A. The directed relationship is indicated by the arrow from node A to node C. Above each network is the adjacency matrix for that graph. You can see that there are differences between the two matrices. For example, the link between node A and node C in the undirected graph shows 1 in row A column C and 1 in row C column A, indicating that there is a link in both directions. In the directed graph, the link flows from node C to node A so the matrix shows 1 in row C column A and 0 in row A column C as the flow is only in one direction.

A graph will appear different depending on how the links are defined. A relationship "boss of" will result in a link with a different direction to the relationship "reports to." Care must be taken in defining and assigning weights to relationships to ensure that accurate results are obtained for the situation under study. Unreliable input will result in unreliable output. You will notice that the diagonal cells across the matrix are each filled with a zero, indicating that there is no link between the node and itself. A link from a node to itself is termed *recursive* and this is not common in social network analysis. If you cut the adjacency matrix for an undirected graph along the main diagonal, one half will be a mirror image of the other half. The values of the relationships or edges for these undirected and directed graphs are binary, either 0 or 1. It is possible, however, for each relationship to be weighted or valued, carrying a larger range of values than just a binary. For example, relationships between nodes could be measured by a ranking between 0 and 5. Take the situation where you are investigating a network of employees and the people they approach for advice in carrying out their job. The link value indicates how often they confer with other colleagues for advice. A ranking of 0 could indicate they did not confer at all with that particular colleague and 5 may mean that they conferred on a daily basis.

Valued relationships can indicate amount by quantification, strength, or frequency. As the value of links change, the network can also change. *Phase transitions* occur when a network moves from one state to another as the weighting or value of links change. For example, as the number or value of links increases in comparison to the number of nodes, the network changes from a collection of disconnected nodes to a connected state. The network moves from one state to another.

Another way to represent network data is an *edge list*. An *edge list* is a compact way to record sparse network data (i.e., networks with few links). The edge list simply lists each link in the network beginning with the source node and ending with the target node. For example, the edge list for the directed network in Figure 1.3b is

A	to	B
A	to	C
C	to	B
C	to	D
E	to	C

The edge list representation is often used by computer scientists to record network data because it takes less memory than recording all of the zeros that may exist in a sparse network. This would not be particularly useful for data structures where all of the potential links had a weighted value, such as a distance network. Many

social scientists find that this is an efficient method to record relational data in interviews, or observational studies as well.

1.4   GOOD WILL HUNTING PROBLEM

Returning to the Good Will Hunting Problem, as it turns out, the adjacency matrix, **A**, for Figure 1.1 is the key to solving the problem. As a check on learning so far, see if you can create the adjacency matrix for Figure 1.1.

Check on Learning

Develop the adjacency matrix for Graph **G** in Figure 1.1.

Answer

$$
\mathbf{A} = \begin{array}{c|cccc} \mathbf{A} & 1 & 2 & 3 & 4 \\ \hline 1 & 0 & 1 & 0 & 1 \\ 2 & 1 & 0 & 2 & 1 \\ 3 & 0 & 2 & 0 & 0 \\ 4 & 1 & 1 & 0 & 0 \end{array}
$$

The Next Step

The next step in solving the Good Will Hunting Problem is finding the number of walks with length 3 present in the graph. Multiple compositions of the adjacency matrix, $\mathbf{A}^n$, gives the number of walks of length n between all nodes in the network. That is, for the adjacency matrix $\mathbf{A}$, $[\mathbf{A}^n]_{ij}$ represents the number walks of length n from node i to node j. Refer to Appendix A, Matrix Algebra Primer, for a review and summary of the matrix algebra steps necessary to perform compositions of the adjacency matrix. Therefore, since

$$
\mathbf{A} = \begin{bmatrix} 0 & 1 & 0 & 1 \\ 1 & 0 & 2 & 1 \\ 0 & 2 & 0 & 0 \\ 1 & 1 & 0 & 0 \end{bmatrix}
$$

then

$$\mathbf{A}^2 = \begin{bmatrix} 0 & 1 & 0 & 1 \\ 1 & 0 & 2 & 1 \\ 0 & 2 & 0 & 0 \\ 1 & 1 & 0 & 0 \end{bmatrix} * \begin{bmatrix} 0 & 1 & 0 & 1 \\ 1 & 0 & 2 & 1 \\ 0 & 2 & 0 & 0 \\ 1 & 1 & 0 & 0 \end{bmatrix} = \begin{bmatrix} 2 & 1 & 2 & 1 \\ 1 & 6 & 0 & 1 \\ 2 & 0 & 4 & 2 \\ 1 & 1 & 2 & 2 \end{bmatrix}$$

and

$$\mathbf{A}^3 = \begin{bmatrix} 2 & 1 & 2 & 1 \\ 1 & 6 & 0 & 1 \\ 2 & 0 & 4 & 2 \\ 1 & 1 & 2 & 2 \end{bmatrix} * \begin{bmatrix} 0 & 1 & 0 & 1 \\ 1 & 0 & 2 & 1 \\ 0 & 2 & 0 & 0 \\ 1 & 1 & 0 & 0 \end{bmatrix} = \begin{bmatrix} 2 & 7 & 2 & 3 \\ 7 & 2 & 12 & 7 \\ 2 & 12 & 0 & 2 \\ 3 & 7 & 2 & 2 \end{bmatrix}$$

The matrix, $\mathbf{A}^3$, reveals that there are in fact seven walks of length 3 from node 1 to node 2. Table 1.5 lists all walks of length 3 from node 1 to node 2 for this simple network. Notice from the matrix A^3 that there are no walks of length 3 for node 3 back to itself, and there are 12 walks of length 3 from node 2 to node 3. Thus, Will Hunting used network science in the first two steps of his four-step solution. Steps three and four use a branch of mathematics known as *functional analysis* and is beyond the scope of this book. A full solution can be found at the Harvard University Mathematical Archive[1].

TABLE 1.5 3-Step Walks from Node 1 to Node 2

		Step 1		Step 2		Step 3
1.	Node 1	$\rightarrow$	Node 2	$\rightarrow$	Node 4	$\rightarrow$	Node 2
2.	Node 1	$\rightarrow$	Node 2	$\rightarrow$	Node 1	$\rightarrow$	Node 2
3.	Node 1	$\rightarrow$	Node 4	$\rightarrow$	Node 1	$\rightarrow$	Node 2
4.	Node 1	$\rightarrow$	Node 2	$\rightarrow$	Node 3	$\rightarrow$	Node 2
5.	Node 1	$\rightarrow$	Node 2	$\rightarrow$	Node 3	$\rightarrow$	Node 2
6.	Node 1	$\rightarrow$	Node 2	$\rightarrow$	Node 3	$\rightarrow$	Node 2
7.	Node 1	$\rightarrow$	Node 2	$\rightarrow$	Node 3	$\rightarrow$	Node 2

[a] The last four walks are not repeats. Since nodes 2 and 3 are in a loop, there are four walks between them. Thus, walk 4 takes the path of going across the top link and back to node 2. Walk 5 takes the path going along the bottom link. Walk 6 uses both links from top to bottom, whereas walk 7 reverses and goes from bottom to top.

A Second Problem

The movie Good Will Hunting proposed a second problem that relates to networks. Professor Gerald Lambeau comes to the class and announces a problem that has taken MIT professors 2 years to solve. He boldly proclaims to the class that the so-called gauntlet has been thrown down.

[1] http://www.math.harvard.edu/archive/21b_fall_03/goodwill/

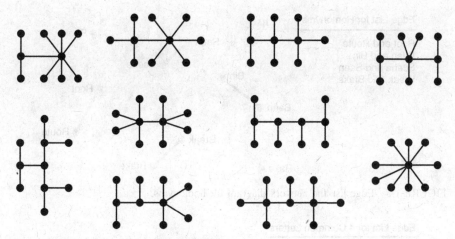

FIGURE 1.4   Good Will Hunting solution.

Only strong students would dare try and solve it. The problem is, *Find all home-omorphically irreducible trees of degree ten (i.e., ten vertices).* In the language of network science, the problem is to find all combinations of fully connected networks with 10 nodes that contain no cycles. Essentially, it means we need to join 10 nodes together such that all nodes are connected to at least one other node. No cycles are allowed and only one link is allowed between any two nodes. Most importantly, no node can have a degree of 2. How many can you create?

Check on Learning

Find all combinations of fully connected networks with 10 nodes that contain no cycles.

Answer

There are 10 such combinations as shown in Figure 1.4. They are listed here. How many did you find? In the movie, Will Hunting only wrote eight of them. Were you smarter than a genius?

Now Let us Turn to Some Examples

Example 1.1

Fun with Words

Given words: Root, Route, Brake, Break, Seen, Scene, Been, Bin, Rake, Reek. Construct a network assuming the relationship is defined by homonyms (words that sound the same). The homonyms are Root and Route, Brake and Break, Seen and

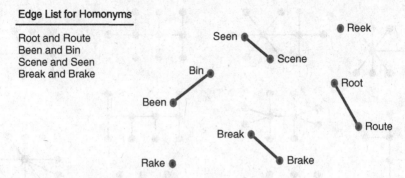

FIGURE 1.5   Edge list and network diagram for homonyms.

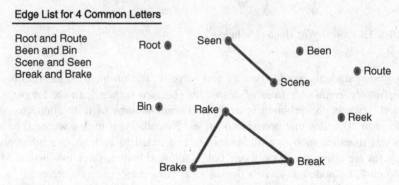

FIGURE 1.6   Edge list and network diagram for four letters in common.

Scene, and Been and Bin. The network can also be represented visually. Figure 1.5 shows a network visualization of the network consisting of the 10 words and the edge link list for the homonyms.

The homonym network has four edges. It is important to observe that we can also define a different relationship between the words. For example, if the relationship is defined as words that have four letters in common, a new edge list and network visualization can be created. This network also has four edges, as seen in Figure 1.6.

Similarly, a relationship of three letters in common will produce the following edge list and network visualization, as illustrated in Figure 1.7.

There are now 10 edges in this network. The important point here is that there is not just one relationship that can be defined on a set of nodes. The same network plotting a relationship of two letters in common will result in a much more connected network, with the edge list and visualization in Figure 1.8.

Look at the network visualizations for two and three letters in common. The network for two letters has 20 edges, whereas the network for three letters has only 10 edges and is fragmented into smaller networks. This is another example of a *phase transition*. The term phase transition comes from physics and represents

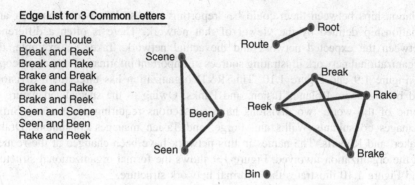

Edge List for 3 Common Letters

Root and Route
Break and Reek
Break and Rake
Brake and Break
Brake and Rake
Brake and Reek
Seen and Scene
Seen and Been
Rake and Reek

FIGURE 1.7   Edge list and network diagram for three letters in common.

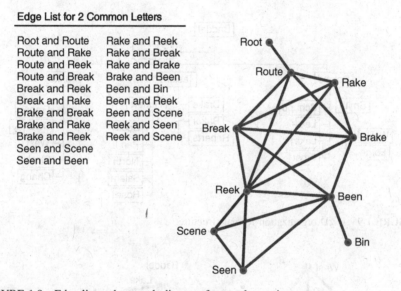

Edge List for 2 Common Letters

Root and Route	Rake and Reek
Route and Rake	Rake and Break
Route and Reek	Rake and Brake
Route and Break	Brake and Been
Break and Reek	Been and Bin
Break and Rake	Been and Reek
Brake and Break	Been and Scene
Brake and Rake	Reek and Seen
Brake and Reek	Reek and Scene
Seen and Scene	
Seen and Been	

FIGURE 1.8   Edge list and network diagram for two letters in common.

the point at which the network will rapidly change from fully connected to many components (or vice versa) as the threshold defining a link changes.

In social networks, friends are often different relationships than that of family, or mentorship, or even respect. A relationship can even be negative such as enemy, dislike, or contempt. The definition of what constitutes a relationship can therefore have a significant impact on the analysis of network data.

1.5   FORMAL AND INFORMAL NETWORKS

Networks of people usually relate to a structured situation or group of interest. In an organizational situation, the network of people will be employees and the relationships between them could be "reporting to," "seeks advice from," or any relationship defined by the viewer of that network. There is often a difference between the expected network and the actual network. Take, for example, the organizational network illustrating sources of important information between people in Figure 1.9 and Figure 1.10. This R&D organization has three main divisions led by managers Roland, Tasien, and Panks. Owing to the specialized nature of some of the work, two divisions have subsections requiring consultants; Smith manages consultants Wallis and Forge, and Tasien manages consultants Drake, Baker, and Roberts. The names in this network have been changed at the request of the organization involved. Figure 1.9 shows the formal organizational structure and Figure 1.10 illustrates the informal network structure.

Most managers are aware of pockets of relationships that exist between their coworkers but many are not aware of the informal network structure of their entire

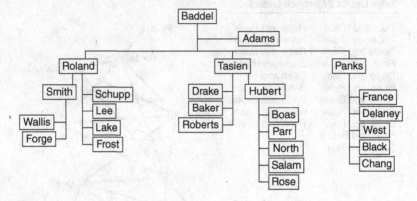

FIGURE 1.9   R&D organization formal structure.

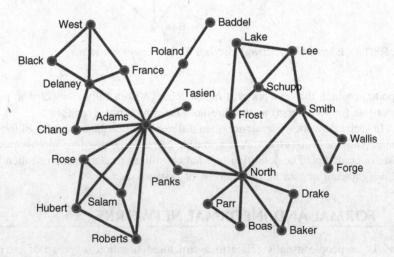

FIGURE 1.10   R&D Organization informal structure.

organization or department. In Figure 1.9, Adams and North do not have prominent positions in the formal organizational network (North being fairly low rank under Hubert's management and Adams being the administrative assistant to Baddel); however, Figure 1.10 shows that both have a central role in the informal organizational structure. North, who is not ambitious, has been employed with the organization for many years and has not sought promotion. He has professional knowledge and experience that surpasses all other employees. Adams holds power due to her access to equipment and funds, and her extensive knowledge of procedures. If resources cannot be obtained through the official channels, Adams knows how to obtain them through other channels. Figure 1.10 clearly shows that the organization consists of two smaller networks, with Adams and North being the gateways between these two networks. A comparison of the formal and informal structures also indicates that very few employees go to their managers for important information. The senior managers appear to be distanced from their employees. Communications and information flows in the informal organization tend to bear little resemblance to that of the formal organization.

Example 1.2

Military Chain of Command and Informal Power

In order to illustrate the importance of understanding social networks, here is an example of an informal network in a military organization. Consider the noncommissioned officer (NCO) support chain in an Army company. In this example organization, the first sergeant (1SG) was new to the company. This was his second assignment as a 1SG. In his previous assignment, he brought some great ideas to his company and made significant improvements. The command sergeant major (CSM) thought that the 1SG would be able to make similar improvements in our example company. Unfortunately, the 1SG's ideas were not working in the same way that it did with the previous company. Good ideas can often fail when they are implemented poorly. Understanding the informal networks will provide an insight into some important organizational dynamics.

Before we look at the informal network, we must understand the formal chain of support network in an Army company. The chain of support is illustrated in Figure 1.11.

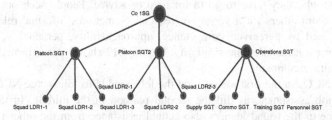

FIGURE 1.11   Formal NCO chain of support.

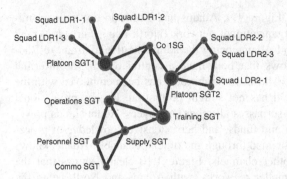

FIGURE 1.12 Informal NCO network.

The 1SG is the senior NCO in the company. His pay grade is E-8 and he has usually served in leadership positions such as squad leader and platoon sergeant (PSG). There are two PSGs in the company as well as an operations sergeant. These NCOs are sergeants first class (SFC), or pay grade E-7. The operations sergeant is sometimes referred to as the *headquarters PSG*. His soldiers are responsible for providing support to the platoons. The SFCs have usually served previously as a squad leader. Their direct supervisor is typically a lieutenant. While the chain of command consists of commissioned officers, the chain of support is a parallel network of NCOs that provide the officers with advice, experience, and logistic support. Each of the platoons has three squad leaders who are staff sergeants (SSGs), pay grade E-6. The company headquarters consists of one SSG who serves as the supply sergeant and three sergeants (SGT) of pay grade E-5. The three SGTs serve as the training sergeant, the personnel sergeant, and the communications sergeant. All the nodes represented in Figure 1.11 are sized according to their rank. Thus, the 1SG is represented by a larger size node than the PSG, which is larger than the squad leader, and so on.

Does the 1SG have the power necessary to make changes in the organization? According to the formal chain of support, he does. However, in our study of this organization, the 1SG is not effective in making change. His ideas are no different than they were in a previous company where he was effective. The difference lies in the informal network of the two companies. Informal networks can be extracted in different ways.

The NCOs in this company were asked who they went to for assistance in how to do their job more effectively. The NCOs were not limited in the number of people they could list. This information represents the informal network. Individuals are not told who they have to go to for help or advice. People seek out this type of support from others for a variety of reasons. Sometimes informal relationships are determined by perceived competence, approachability, personality, common interests, or even racial/ethnic similarity. Figure 1.12 shows this informal network for our military company.

The NCOs in the first platoon, on the left, tended to follow the NCO support chain for job assistance, as did the second platoon in the middle. In the second platoon, one of the squad leaders also sought assistance from the other two squad leaders in the platoon. Perhaps this squad leader was new and looked to his peers for mentorship. Members in the headquarters platoon had a similar dynamic. The node that stands out as unusual in the informal network is the training sergeant, who consulted more than any other individual for assistance. Even the 1SG looks to the training sergeant for assistance.

There are several reasons for this informal organizational structure. The training sergeant had been serving in this capacity for almost a year and a half, which is a long time to serve in a duty position in the military. The training sergeant knew almost all of the NCOs who served in the higher battalion headquarters. He would often spend off-duty time with the battalion NCOs as well as with NCOs in his own company. In addition, he belonged to a couple of social groups that also had NCOs different companies as members. Through this larger social network, the training sergeant was able to more effectively coordinate training resources through his friendship ties at the battalion level than other NCOs in other companies. The training sergeant was also able to mobilize his social network to find opportunities for squad leaders to send their soldiers to weapons ranges conducted by other units, use sections of training area reserved by other units, and coordinate similar training resources. The training sergeant enjoyed his position in the company. He liked the fact that senior NCOs would come to him for support. This position in the informal network gave the training sergeant power.

When the 1SG first arrived at the company, he unknowingly hurt his effectiveness in the informal network. He wanted to assert himself as a leader and decided that he needed to make sure that soldiers maintained a high standard of military appearance and bearing. The training sergeant did not make a good impression on the 1SG. The 1SG felt that the training sergeant did not maintain a professional appearance and that he was cavalier and borderline disrespectful. This may not be entirely surprising, considering the power that the training sergeant held in the informal network. The 1SG pointed out several corrections to the training sergeant and expressed his concerns about military discipline to the operations sergeant. The training sergeant was insulted and embarrassed by this first meeting. Yet, he did not openly discuss his feelings. As a result, the training sergeant was not eager to help make the new 1SG successful, nor was he enthusiastic about implementing any of the 1SGs new ideas. Further, the training sergeant disagreed with many of the 1SGs ideas. To make the company more efficient, the 1SG required NCOs to brief him on resources they were considering for their training. He would then balance those resources across the company. This of course, took away some of the power that the training sergeant enjoyed.

After a few months, it was time for the training sergeant to conduct a permanent change of station and a new NCO assumed the duties of training sergeant and the 1SGs ideas slowly made the company better. However, these improvements might have been adopted more rapidly and more effectively had the 1SG been aware of the informal social network in the company. Perhaps, if the company leadership had been able to monitor the social network, they would have prevented a relatively junior NCO from having so much power in the organization. Perhaps, the company leadership could have utilized the training sergeants informal network through appropriate incentives. In any case, this example illustrates the importance of understanding the informal social network in an organization.

Social network analysis offers more than pictures: it provides an entirely new dimension of statistical analysis for organizational behavior. Traditional analysis focuses on individual attributes. Social networks focus on relationships between individuals. Traditional analysis assumes statistical independence, whereas social network analysis focuses on dependent observations. Traditional analysis seeks to identify correlation between significant factors and a response variable. Social network analysis seeks to identify organizational structure. The underlying mathematics behind traditional analysis is calculus, the language of change. The corresponding mathematics behind social network analysis is linear algebra (Lay, 2005) and graph theory (West, 2000). These differences can be significant in terms of how someone looks at an organization.

1.6 SUMMARY

Social network analysis provides the tools to analyze situations under study as networks, provided nodes can represent objects or individuals, and links or edges denote relationships. Network analysis has the advantage of being able to analyze entire networks such as organizations, as opposed to traditional analysis that concentrates on analyzing the significance of individual factors. In an organizational setting, network analysis enables the analyst to study the links between people, which can indicate both positive and negative relationships. By studying both the formal reporting structures as well as the informal social networks, analysts can provide valuable information to decision makers within the organization.

Here is a summary of the key points discussed in this chapter:

- The key components needed to analyze and visualize a network are the adjacency matrix and the graph.

- Almost anything can be represented as a node or vertex. A node in a network could represent people, knowledge, roles, beliefs, or any other object of interest.

- Different relationships can be defined on a given set of nodes or vertices. Relationships can be represented by a binary yes/no, 0/1; or a number or code. Relationships between nodes can take the form of associations, ties, and attachments that connect one node to another.

- Relationships can be undirected/reciprocal or directed, indicating the direction of the link.

- Network structure will appear very different depending on how relationships are defined.

- Where links between nodes in a network carry values, phase transitions can occur when valued relationships move over the threshold that either fragments the network into smaller segments, or combines previously unconnected nodes into more coherent network structures.

CHAPTER 1   LAB EXERCISE

Introduction to Networks and Using the Organizational Risk Analyzer (ORA) Software System

Learning Objectives

1. Play the Kevin Bacon game.
2. Learn how to start using ORA.
3. Learn how to load a meta-network.
4. Learn how to produce a visualization of the meta-network using ORA.

1. Play the Kevin Bacon Game

Go to the following URL and enter the name of a movie star to see how many links away from Kevin Bacon.

http://oracleofbacon.org/

Have you heard of six degrees of separation? What does this mean? Google it!

2. Load the ORA Software

Load the ORA software from the CASOS website (URL below). ORA is free, but you need to register, so please enter your email address.

http://www.casos.cs.cmu.edu/projects/ora/software.php

You will find a variety of User Guides and Tutorials for download from the CASOS web site at

http://www.casos.cs.cmu.edu/projects/ora/publications.php

3. Lab Exercise 1 Network

You will be using the LabExercise1data provided in the following table:

Using ORA: You can create a new meta-network in ORA or load previously defined meta-networks. Meta-networks comprise nodes and links. You must create a meta-network, then add node classes, and then you will be able to add links between nodes to form networks.

4. Start ORA

Double click on the ORA icon to start ORA running

The main ORA screen will appear:

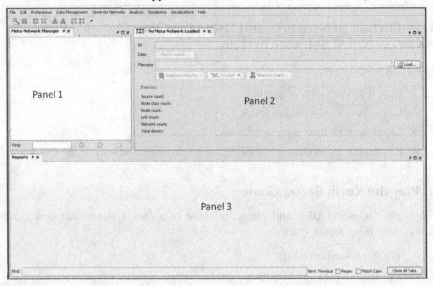

Panel 1 displays a list of the meta-networks that has been loaded into ORA, its components, and sub-networks. To view information about a particular meta-network or sub-network, single-click on the meta-network name.

Panel 2 displays information about the meta-network or sub-network highlighted in Panel 1.

Panel 3 displays information about the results of analyses run on your meta-network.

We will now create a new meta-network called *LabExercise1* and enter the data in the tables for the meta-network named Lab Exercise 1 (data tables can be found in Appendix B). This meta-network has only one node class named Agent, and one network named Agent × Agent.

Let us begin. Start ORA by double clicking on the ORA icon. You should get the introductory screen above. Select the icon for a new meta-network.

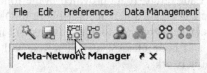

ORA will create a meta-network in Panel 1 called *New Meta-Network*

In Panel 2 enter the network ID as LabExercise1 (replacing New Meta-Network) and press Click to Create.

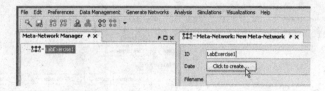

We will now add the node class Agent to the meta-network. Select the icon for a new node class.

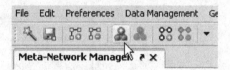

Leave the node type and ID as Agent.

Enter 10 for the Size field in the Create Node Class dialog box to generate 10 Agent nodes. Then select Create.

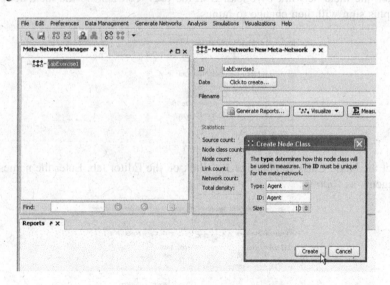

We will now create an Agent × Agent network. Select the icon for a new network.

Enter the name Agent × Agent for the relationship in the dialog box. Select Create.

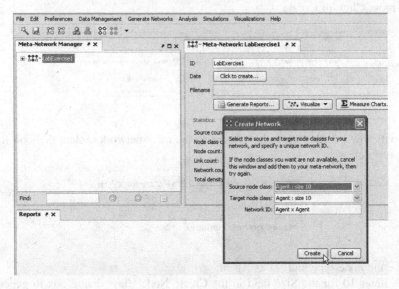

Expand the meta-network by clicking on the plus icon next to the meta-network. The plus sign will then change to minus.

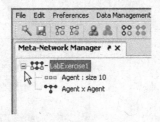

Select the Node Class Agent: size 10. Click on the Editor tab. Enter the names for the agents as follows:

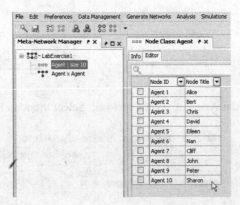

Select the network Agent × Agent; select the Editor tab in the editor window to the right; and double click on a cell to insert a tick. A tick represents 1 in the adjacency matrix for Lab Exercise 1 and no tick represents 0.

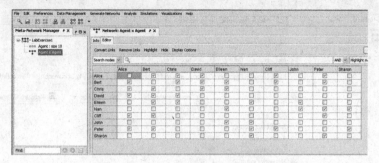

Use the Display Options tab to toggle between binary and numeric link display. What does each option do? Try them and see.

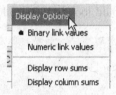

We must save our work before we go any further. Save the meta-network by selecting File → Save As. Name the meta-network LabExercise1. It will be saved as an .xml file named LabExercise1.xml.

When we want to open this meta-network again, select File → Open Meta-Network

Click on the Meta-Network *LabExercise1* again then choose the *Visualize* tab.

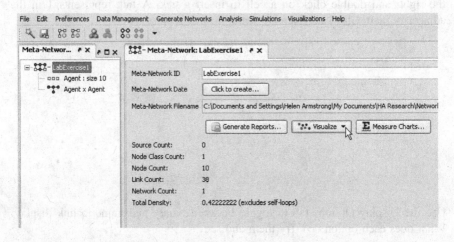

Your network visualization should look like the following:

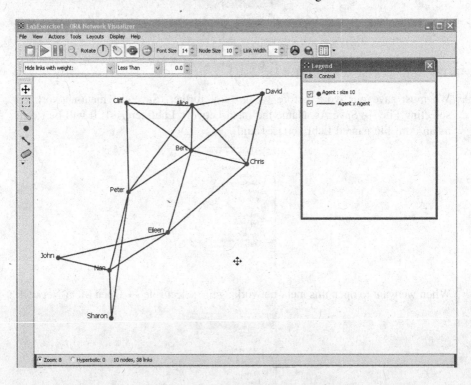

You can toggle to change the background color between white and black by:

→ Display → Background color → Black

So that you can see the network in full, move the Legend box by clicking on the blue strip at the top and drag it.

Play with → Display options → Node Appearance and Background color to change the background to other colors, change the color and size of the nodes, etc.

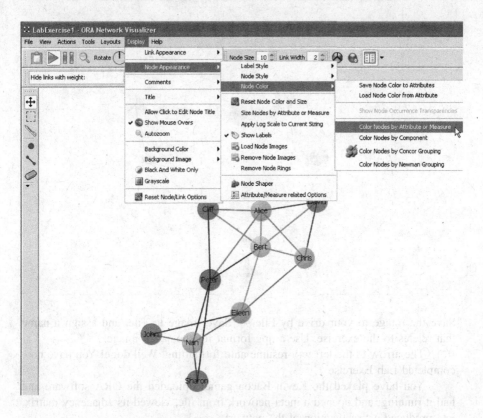

By pressing the Stop/Pause layout button on the top left of the screen, you can stop the automatic formatting and reposition the nodes.

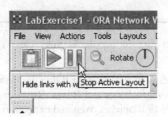

Click on a node, hold it, and drag it to where you want it positioned, then release. You will notice that the lines attached to the node will move with the node. Keep repositioning nodes until you are happy with the diagram.

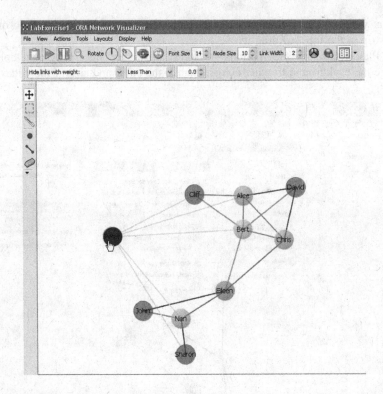

Save the image to your drive by File → Save Image to File, and assign a name that relates to the exercise. Use a .jpg format for the saved image.

The arrow to the left will resume auto formatting. Well done! You have now completed Lab Exercise 1.

You have played the Kevin Bacon game; uploaded the ORA software and had it running; and opened a meta-network from file, viewed its adjacency matrix, and produced a visualization of the network.

EXERCISES

1.1. Construct a social network of people in your class or of those known to the class.

1.2. Construct a directed graph to illustrate the following network of countries that trade together. Cameria exports to Puma and Rovia. Rovia exports to Gundi, Isicca, and Wenland. Puma exports to Klapistan. Wenland exports to Promany, Nigaly, Englore, and Isicca. Nigaly exports to Isicca and Gundi.

1.3. Construct the binary adjacency matrix for the network in Exercise 1.2.

1.4. Construct a directional network to represent the following situation. Make sure you show the direction of the relationship.

Employee	Who they go to for advice
Bill	Fred, Heather, Paul
Fred	Mary, Paul, Julie, Heather
Mary	Julie
Gordon	Heather
Heather	Paul
Julie	Paul, Heather
Paul	Heather

1.5. Construct the binary adjacency matrix for the network in Exercise 1.4.

1.6. Using an automated network analyses and visualization tool, enter the adjacency matrix for the data in the friendship graph (Fig. 1.2) and produce an image of the network using the visualization function of the tool.

1.7. Using an automated network analyses and visualization tool, complete the adjacency matrix for the informal organizational structure for the friendship graph (Fig. 1.2) and produce an image of the network using the visualization function of the tool.

REFERENCES

Lay, D. (2005). *Linear Algebra and Its Applications*, 3rd updated edition. Addison Wesley.
West, D. (2000). *Introduction to Graph Theory*, 2nd edition. Prentice Hall.

CENTRALITY MEASURES

> The centrality of group effort to human life means that anything that changes the way groups function will have profound ramifications for everything from commerce and government to media and religion.
>
> —*Clay Shirky*, Here Comes Everybody

Learning Objectives

1. Understand the importance of measuring centrality.
2. Explain the centrality measures: degree, closeness, betweenness, and eigenvector.
3. Calculate degree, closeness, and betweenness centrality for a network.
4. Explain the effect of directed versus undirected data on the measures.
5. Understand the difference between random, small world, core-periphery, and scale-free network topologies.

2.1 WHAT IS "CENTRALITY" AND WHY DO WE STUDY IT?

When we look at social networks, one natural question to ask would be who is most important in the network? Figure 2.1, which we call a *core-periphery* graph, contains five nodes in the "core" (Agents 1, 3, 15, 24, and 28) and the rest of the nodes in the "periphery." We have sized the nodes by the number of links they have to highlight the core members of the graph.

What if the graph is not as neatly partitioned into core and periphery? The network depicted in Figure 2.2 is not very neat. If we size the nodes according to their number of links we gain little insight apart from the observation that Agents 23, 24, and 26 have more links than the others as shown in Figure 2.3a.

This process of counting how many links each agent has measures what we call *degree centrality*. An agent's degree is simply the count of the number of links going into it, *in-degree*, or coming from it, *out-degree*. This centrality measure is easy to comprehend. But, as we just saw, degree centrality is a bit limiting.

Social Network Analysis: with Applications, First Edition.
Ian A. McCulloh, Helen L. Armstrong, and Anthony N. Johnson.
© 2013 John Wiley & Sons, Inc. Published 2013 by John Wiley & Sons, Inc.

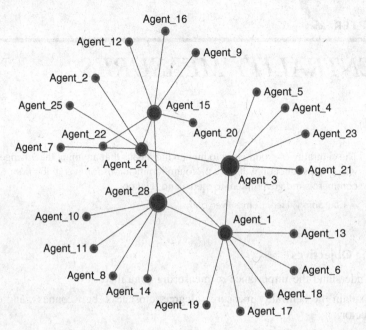

FIGURE 2.1   Core periphery graph.

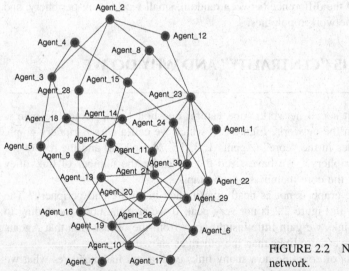

FIGURE 2.2   Not-so-neat network.

How do we determine which other agents are important in this network? Perhaps we would like to know who is connected to important agents (i.e., agents with many links). We call this particular measure *eigenvector centrality*. Figure 2.3b is the same graph as Figure 2.3a, but this time each agent is sized according to its eigenvector centrality. We can see a difference in some of the agents: a closer look

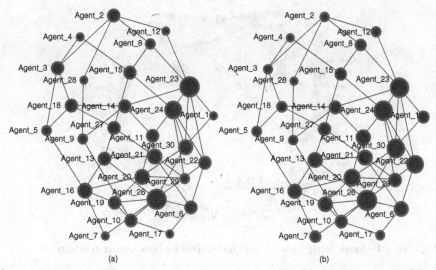

FIGURE 2.3   Comparing graphs with nodes sized by (a) degree centrality and (b) eigenvector centrality.

at Agents 22, 29, and 30 shows a greater centrality when measured by eigenvector centrality.

There may be other important characteristics we wish to discover. Consider a social network of employees within a large corporation. There are people we might traditionally think of as important, such as the chief executive officer (CEO), chief operating officer (COO), or chief financial officer (CFO). Intuition leads us to believe that these individuals will have a high degree centrality as they are the most important members of the company. But, in practice, they tend to have low degree centrality. Why? Because they are connected to few other members, namely, secretaries and middle managers, who actually execute their demands for them. The secretaries and middle managers now form a "chokepoint" so that the social network resembles the core-periphery network of Figure 2.1. In social network analysis, we would say that the secretaries and middle managers have high *betweenness centrality* as they lie along many of the communication lines between people in the corporation. Further, the middle managers are high in *closeness centrality* as they require the fewest number of steps to many people in the company.

Figure 2.4 is a *small world network* as studied by Watts and Strogatz (1998). It contains a few *spanners*. Spanners form connections across the organization and have high betweenness centrality. In Figure 2.4, the nodes have been sized on the basis of their betweenness centrality. It is clear that the spanners have very high betweenness centrality; this is because, in order for one agent to quickly get to another agent on the other side of the "small world," the path often includes spanners.

Figure 2.5 is a *scale-free* network studied extensively by Barabási and Albert (1999). It contains a few agents with many connections, and many agents with few connections—much like a network of web references among websites. These agents

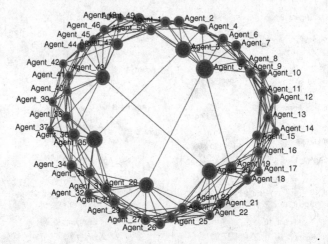

FIGURE 2.4   Small world network with nodes sized by betweenness centrality.

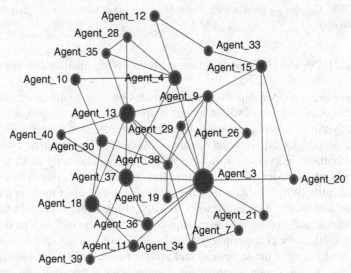

FIGURE 2.5   A scale-free network with nodes sized by degree centrality, and isolate nodes are hidden.

with many connections will be very close to many other agents, as we can see in Figure 2.5. The nodes are sized by degree centrality, and nodes with many links are often referred to as *hubs* in scale-free networks. In a network of websites, hub nodes would be Google, Yahoo, Ebay, Amazon, and similar large web hubs.

Check on Learning

In Figure 2.4, which nodes would you identify as *spanners*? What special property do these particular nodes have in common?

Answer

Spanners are those large nodes situated in the inner section of the graph. These spanner nodes are high in betweenness centrality and connect to others across the network, spanning the hole in the middle of the graph.

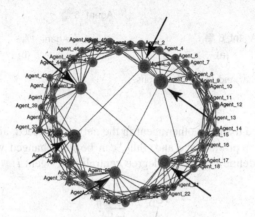

2.2   CALCULATING NODAL CENTRALITY MEASURES

In calculating centralities let us review once again some simple definitions. A network, also called a *graph* in mathematics, is made up of points, usually called *nodes* or *vertices*, and the lines connecting them, usually called *links* or *edges*. We will now explain how to calculate degree centrality, betweenness centrality, closeness centrality, and eigenvector centrality measures at the node level based on the seminal work of Freeman (1979) and Bonacich (1972).

In the calculation of all centrality measures, the total number of nodes in a network is n, and the total number of possible links a node may have to other nodes will be $(n - 1)$ as nodes in social networks seldom link back to themselves. For example, as there are 40 agent nodes in Figure 2.5, i.e. $n = 40$, the total number of possible links a node may have will be $(n - 1)$, which is 39.

2.2.1   Degree Centrality

The simplest definition of nodal centrality is the assumption that nodes or agents with the most ties to other agents must hold a special place of influence within the network or graph. Probably the best place to see this is when we analyze the star graph in Figure 2.6a. In this type of graph it is easy to see that one agent, Agent_4 in this case, has ties to all $n - 1$ other agents and the remaining agents have their single connection to the network through Agent_4. This gives Agent_4 the highest centrality index when the measure is the total number of connections an agent has to other agents. This is degree centrality. Now contrast the star graph, Figure 2.6a, with the circle graph, Figure 2.6b. In the circle graph, there is no one agent more

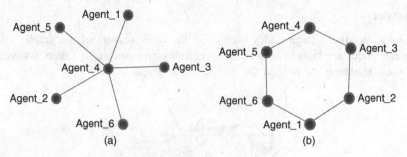

FIGURE 2.6   (a) Star and (b) circle graphs.

active or connected than any other agent in the network. Indeed, all the agents have the same degree or prominence and, thus, can be interchanged with no effect on the network. We define C_{Di} as the degree centrality of agent i.

$$C_{Di} = \sum_{j=1}^{n} a_{ij}$$

The degree centrality is equal to the sum of all of the links connected to an agent. From Table 2.1, the adjacency matrix $\mathbf{A}$, this equates to summing across the rows. We can, further, divide by $n - 1$ to standardize the measure so that it can be compared with other networks of varying sizes of n.

$$C_{Di}^* = \frac{\sum_{j=1}^{n} a_{ij}}{(n-1)}$$

Example 2.1

We wish to determine the degree centrality for Agent_4 in the star graph in Figure 2.6a. Using the adjacency matrix, Table 2.1, we simply sum the row for Agent_4. Note that diagonal entries are zero because our graphs or networks are nonreflexive (a node not connected to itself). So then to calculate the degree of Agent_4, we simply do the following:

$$C_{D4}^* = \frac{\sum_{j=1}^{6} a_{4j}}{(6-1)} = \frac{1+1+1+0+1+1}{5} = \frac{5}{5} = 1$$

2.2.2 Betweenness Centrality

The next centrality measure we examine is betweenness, where C_{Bi} is defined as the betweenness centrality of agent i. To talk about betweenness, we must first discuss the concepts of paths and geodesics. As mentioned previously, a path in

TABLE 2.1   Star Graph Matrix

	Agent_1	Agent_2	Agent_3	Agent_4	Agent_5	Agent_6	Sum total
Agent_1	0	0	0	1	0	0	1
Agent_2	0	0	0	1	0	0	1
Agent_3	0	0	0	1	0	0	1
Agent_4	1	1	1	0	1	1	5
Agent_5	0	0	0	1	0	0	1
Agent_6	0	0	0	1	0	0	1

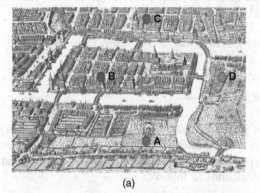

(a)

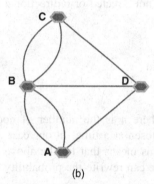

(b)

FIGURE 2.7   Euler bridge problem: (a) map of Konigsberg and (b) graph of Konigsberg bridges.

a network is simply a listing of nodes that are sequentially connected and distinct (nodes and links are not used more than once).

In the 1700s the city of Koenigsberg, Prussia, was built on the Pregel River with its land masses connected by seven bridges, as shown in Figure 2.7a. The mathematician Leonard Euler studied whether it was possible to find a path that crosses all seven bridges only once, and he attempted to find paths from one starting location to another. He could never cross the same area of the city or land mass (node) more than once, nor could he cross the same bridge (link) more than once. Figure 2.7b shows the situation as a network, with the nodes representing the land masses and the links are the bridges. Finding a way through Koenigsberg that involved crossing each bridge only once was a real challenge. The bridges were the only means of reaching the islands and each bridge must be crossed completely each time it was entered. Walking halfway on to the bridge and then turning around and approaching the same bridge from the other side was not permitted. Euler was unable to deliver a solution to the problem.

To calculate the betweenness centrality measure for the simple graph in Figure 2.8, we need to study the paths in the network. {1,2} is a path, as is {1,2,5}, as they never cross the same link or agent more than once. {1,2,4,2,5}, on the other hand, is not a path, as it crosses Agent_2 twice. A *geodesic*, defined as g_{ij}, is the shortest path between agent i and agent j. For example, in the graph above, the geodesic between Agent_1 and Agent_4 is {1,2,4}. Another way of saying this is that links l_{12} and l_{24} forms the path between Agents 1 and 4. The term *diameter* refers to the longest geodesic in the network. To calculate the betweenness centrality, first find all of the geodesies in the graph or network. This means that you must determine the total number of dichotomous paths present. This will be used to examine the geodesics between the two nodes. When we say *dichotomous path*, we mean by implication that there are exactly two nodes in view and the path between them. We, further, assume that the paths are nonreflexive. The total number of possible dichotomous paths turns out to be either the familiar probability calculation for permutations (nPr), if the network is directed and order matters, or the equally familiar probability calculation for combinations (nCr), if the network is not directed or bidirectional and order does not matter. They are calculated as

$$nPr = \frac{n!}{(n-r)!}$$

and

$$nCr = \frac{n!}{r!(n-r)!}$$

where n is the number of nodes in the network and r is the number of nodes chosen at a time. In our case, we are interested in links between pairs of nodes. This means that for the above calculations r will always be 2. With that in mind, we can rewrite the probability calculations as follows.

$$nP_2 = \frac{n!}{(n-2)!} = \frac{n(n-1)(n-2)!}{(n-2)!} = n(n-1)$$

and

$$nC_2 = \frac{n!}{2!(n-2)!} = \frac{n(n-1)(n-2)!}{2!(n-2)!} = \frac{n(n-1)}{2}$$

So, for the undirected network in Figure 2.8, the total number of dichotomous paths to consider would be

$$_5C_2 = \frac{5(5-1)}{2} = \frac{5(4)}{2} = \frac{20}{2} = 10$$

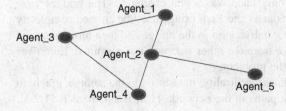

FIGURE 2.8 Example graph for betweenness calculation.

TABLE 2.2 List of Geodesic Paths

Path number	From node	To node	Geodesic path(s)
1	1	2	{1,2}
2	1	3	{1,3}
3	1	4	{1,2,4};{1,3,4}
4	1	5	{1,2,5}
5	2	1	Same path as (1,2)
6	2	3	{2,1,3};{2,4,3}
7	2	4	{2,4}
8	2	5	{2,5}
9	3	1	Same path as (1,3)
10	3	2	Same path as (2,3)
11	3	4	{3,4}
12	3	5	{3,4,2,5};{3,1,2,5}
13	4	1	Same path as (1,4)
14	4	2	Same path as (2,4)
15	4	3	Same path as (3,4)
16	4	5	{4,2,5}
17	5	1	Same path as (1,5)
18	5	2	Same path as (2,5)
19	5	3	Same path as (3,5)
20	5	4	Same path as (4,5)

The above calculation tells us that we must examine 10 separate dichotomous paths between pairs of nodes. Table 2.2 lists the geodesics associated with each pair. Note that some node pairs have more than one geodesic path.

The betweenness centrality measure can be standardized. C_B^* would then have both a numerator and a denominator. When looking at the 13 geodesics between the 10 dichotomous paths of Table 2.2, every time a node n lies on a geodesic, add one or a fraction thereof to its betweenness score. This is the numerator of the standardized measure, which can be defined mathematically as

$$C_{Bi} = \frac{\sum_{i<j} g_{jk}(\text{Agent\_}i)}{g_{jk}}$$

where g_{jk} represents the number of geodesic paths between Agent_j and Agent_k, and $g_{jk}(\text{Agent\_}i)$ represents the number of geodesic paths between Agent_j and Agent_k that contain Agent_i. Table 2.3 is an example of how we calculate the betweenness of Agent_2 of Figure 2.8. In some instances a node does not have an exclusive geodesic. For example, between Agent_1 and Agent_4 there are two geodesics {1,2,4} and {1,3,4}, and between Agent_3 and Agent_5 there are also two geodesics {3,4,2,5} and {3,1,2,5}. In this situation, each geodesic is apportioned

TABLE 2.3 Geodesic Path Calculation for Numerator of Agent_2

Path number	From node	To node	Geodesic path(s)	Agent_2
1	1	2	{1,2}	0
2	1	3	{1,3}	0
3	1	4	{1,2,4};{1,3,4}	0.5
4	1	5	{1,2,5}	1
5	2	1	Same path as (1,2)	
6	2	3	{2,1,3};{2,4,3}	0
7	2	4	{2,4}	0
8	2	5	{2,5}	0
9	3	1	Same path as (1,3)	
10	3	2	Same path as (2,3)	
11	3	4	{3,4}	0
12	3	5	{3,4,2,5};{3,1,2,5}	1
13	4	1	Same path as (1,4)	
14	4	2	Same path as (2,4)	
15	4	3	Same path as (3,4)	
16	4	5	{4,2,5}	1
17	5	1	Same path as (1,5)	
18	5	2	Same path as (2,5)	
19	5	3	Same path as (3,5)	
20	5	4	Same path as (4,5)	
			Total Sum ⇒	3.5

a fraction, and in the case of geodesic paths {1,2,4} and {1,3,4} since Agent_2 lies in only one of the two paths, 0.5 is added to the betweenness score. Where three geodesics occur between two nodes, each would carry one-third or 0.333. Betweenness is accumulated only when the node being analyzed falls along the geodesic of two other nodes and is not the initial or final node in the geodesic. The score for geodesic number 12 in Table 2.2 (from Agent_3 to Agent_5) would then be $0.5 + 0.5 = 1$, as Agent_2 is between the other nodes on both paths.

For the denominator, we calculate the number of pairs of agents which do not include n. So we are either permuting or combining $(n - 1)$ chosen two at a time. Again, we utilize our probability calculations so that

$$_{(n-1)}P_2 = \frac{(n - 1)!}{(n - 1 - 2)!} = \frac{(n - 1)(n - 2)(n - 3)!}{(n - 3)!} = (n - 1)(n - 2)$$

and

$$_{(n-1)}C_2 = \frac{(n - 1)!}{2!(n - 1 - 2)!} = \frac{(n - 1)(n - 2)(n - 3)!}{2!(n - 3)!} = \frac{(n - 1)(n - 2)}{2}.$$

The standardized measure for betweenness centrality becomes

$$C_{Bi}^* = \frac{C_{Bi}}{(n-1)C_2}$$

Check on Learning

What is the total number of possible links for a directed network with 50 nodes?

Answer

Use a *permutation* as follows:

$$_{50}P_2 = 50(50-1) = 50(49) = 2450$$

Check on Learning

What is the total number of possible links for an undirected network with 50 nodes?

Answer

Use a *combination* as follows:

$$_{50}C_2 = \frac{50(50-1)}{2} = \frac{50(49)}{2} = \frac{2450}{2} = 1225$$

Example 2.2

Let us find the betweenness centrality of Agent_2. From Table 2.3, C_{Bi} is calculated to be 3.5. Next, for $_{(n-1)}C_2$ we use

$$_{(n-1)}C_2 = \frac{(n-1)(n-2)}{2} = \frac{(5-1)(5-2)}{2} = \frac{4(3)}{2} = 6$$

Therefore,

$$C_{B2}^* = \frac{C_{Bi}}{(n-2)(n-1)/2} = \frac{3.5}{6} = 0.5834$$

is the betweenness centrality of Agent_2.

2.2.3   Closeness Centrality

We now turn to the closeness centrality measure defined as

$$C_{Ci} = \left[\sum_{j=1}^{n} d(n_i, n_j) \right]^{-1}$$

where $d(n_i, n_j)$ represents the geodesic distance between nodes i and j and will be used to measure the closeness of agent i relative to all other agents in the network. C_{Ci} will reach a maximum of $(n-1)^{-1}$ when one actor is adjacent to all other actors in the network. It is at a minimum when one node in the network is not *reachable*. A node is reachable if there exists a path linking the two nodes. When no path exists between two nodes n_i and n_j, the distance $d(n_i, n_j) = \infty$, making the measure meaningful only when applied to a connected network. This drawback should be considered when using the measure. Closeness is standardized by multiplying it by $(n - 1)$.

$$C_{Ci}^* = \frac{(n - 1)}{\sum_{j=1}^{n} d(n_i, n_j)}$$

The result is an index that ranges between 0 and 1. The inverse of the standardized measure C_{Ci}^* is the *average path length*. Calculating the closeness centrality of a node, then, becomes straightforward. We use the geodesics of the network to find the *average path lengths* for every pair of nodes.

Example 2.3

Calculating closeness requires a look back at Table 2.2, which is the list of geodesic paths depicting the example graph of Figure 2.8. For each agent, we calculate the *average path length* to every other agent. For example, Agent_4 has geodesics: {1,2,4}, {1,3,4}, {2,4}, {3,4}, and {4,2,5} (with lengths 2, 2, 1, 1, and 2, respectively). Therefore, its average path length is

$$\frac{(2 + 1 + 1 + 2)}{(n - 1)} = \frac{6}{4} = \frac{3}{2}$$

Note that the path length from Agent_1 to Agent_4 is 2 even though there are two geodesics, {1,2,4} and {1,3,4}. Notice, also, that lower numbers of this average path length indicate that someone is closer to the other agents in the network. As the lower average path length means higher closeness, we simply take the inverse of this average, and we have the closeness centrality $C_{C4}^* = \frac{1}{(3/2)} = \frac{2}{3} = 0.667$. Higher numbers of this quantity mean very central nodes.

2.2.4 Eigenvector Centrality

A node is high in eigenvector centrality if it is connected to many other nodes who are themselves well connected. This means that a node's eigenvector centrality is dependent on the centrality of adjacent nodes to which it connects. Because of their connectedness to other well-connected people, nodes in social networks with high eigenvectors are considered to be influential nodes in the network. Computing eigenvector centrality involves linear algebra and matrix theory. A primer for the mathematics used to calculate it is located in Appendix A.

To calculate eigenvector centrality, let $\mathbf{x} = x_i$ be the eigenvector centrality of the ith node of a network, and let $\mathbf{A} = [a_{ij}]$ be the adjacency matrix. Then for the ith node in the network, the eigenvector centrality will be the score proportional to the sum of all nodes connected to it. Mathematically, we say

$$x_i = k \sum_{j=C_i} x_j = k \sum_{j=1}^{N} a_{ij} x_j$$

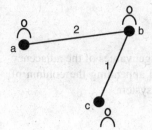

FIGURE 2.9   Weighted graph.

where k is the proportionality constant and C_i is the set of nodes connected to the ith node (x_i). With the judicious choice of $\frac{1}{\lambda} = k$, we have

$$x_i = \frac{1}{\lambda}\sum_{j=1}^{N} a_{ij}x_j$$

which, when rewritten in vector notation, yields the familiar eigenvalue problem.

$$\mathbf{x} = \frac{1}{\lambda}\mathbf{Ax} \quad \text{or} \quad \mathbf{Ax} = \lambda\mathbf{x}$$

Hence, the name *eigenvector centrality*. Typically, there are many values for λ that solve the eigenvalue problem. So an additional requirement is that all values of the solution vector, $\mathbf{x}$, be positive as they relate the scores of individual nodes. Further, necessitated by the Perron–Frobenius theorem (Frobenius, 1912) we desire only the greatest eigenvalue, λ. With the above requirements satisfied, the ith element of the related eigenvector, $\mathbf{x}$, gives the eigenvalue centrality of the ith node in the network.

Example 2.4

Suppose we have the weighted network depicted in Figure 2.9. The adjacency matrix, $\mathbf{A}$, would be

$$\mathbf{A} = \begin{array}{c|c|c|c} & a & b & c \\ \hline a & 0 & 2 & 0 \\ \hline b & 2 & 0 & 1 \\ \hline c & 0 & 1 & 0 \end{array}$$

with column a, b, and c corresponding to nodes a, b, and c, respectively. Useful information about the eigenvalues of the adjacency matrix is encoded in the *characteristic equation* of $\mathbf{A}$. We define the scalar characteristic equation for finding the eigenvalues of $\mathbf{A}$ using the determinant as

$$\det(\mathbf{A} - \lambda\mathbf{I}) = 0$$

where $\mathbf{I}$ is the $n \times n$ identity matrix. The scalar relation

$$\det(\mathbf{A} - \lambda\mathbf{I}) = \begin{bmatrix} -\lambda & 2 & 0 \\ 2 & -\lambda & 1 \\ 0 & 1 & -\lambda \end{bmatrix} = 0$$

yields a polynomial equation of λ,

$$5\lambda - \lambda^3 = 0$$

whose roots, $\lambda_1 = 0$, $\lambda_2 = -\sqrt{5}$, and $\lambda_3 = \sqrt{5}$ are the eigenvalues of the adjacency matrix. Choosing the greatest eigenvalue, λ_3 for λ, and appending the column of zeros gives the rank-deficient augmented matrix of the system

$$(\mathbf{A} - \sqrt{5}\ \mathbf{I})\mathbf{x} = \mathbf{0}$$

which is

$$\begin{bmatrix} -\sqrt{5} & 2 & 0 & 0 \\ 2 & -\sqrt{5} & 1 & 0 \\ 0 & 1 & -\sqrt{5} & 0 \end{bmatrix}$$

From here we can find a basis for the eigenvectors by elementary row reductions. This reveals an equivalent matrix,

$$\begin{bmatrix} 1 & 0 & -2 & 0 \\ 0 & 1 & -\sqrt{5} & 0 \\ 0 & 0 & 0 & 0 \end{bmatrix}$$

which gives the eigenvector

$$\mathbf{x} = \begin{bmatrix} a \\ b \\ c \end{bmatrix} = \begin{bmatrix} 2 \\ \sqrt{5} \\ 1 \end{bmatrix} c$$

where c can be any arbitrary scalar. The parameter, c, is chosen so that the highest eigenvector centrality is 1. So for $c = \frac{1}{\sqrt{5}}$, the eigenvector contains the eigenvector centralities:

$$\mathbf{x} = \begin{bmatrix} a \\ b \\ c \end{bmatrix} = \begin{bmatrix} 0.894 \\ 1.000 \\ 0.447 \end{bmatrix}$$

2.2.5   Google PageRank: A Variant of Eigenvector Centrality

Picture a dozen aircraft hangars filled with documents. Each document contains some information and can be distinguished on the basis of its content. There is no centralized organization of the documents and no catalog system. Further, anyone may add one or more documents at any time without any central logging of its content or record of its existence. The documents simply become another piece of paper in one of twelve enormous hangars full of papers. You stand on a ridge and survey the hangars on the flat land below. Your problem is that you need a vital piece of information. You are certain that the information you need is housed in one of the hangars. You need this information within a matter of minutes, not days or years. How would you go about finding it?

On the surface this seems like a daunting, maybe even an impossible task. Yet, solutions to these types of problems are necessary to deciphering the World Wide Web, a huge, highly disorganized collection of web page documents in various formats. Search engines are obsessed with finding ways to parse and organize the mountains of information into usable morsels able to be categorized. They retrieve pages from the web, index the words in each document, and store the information in an efficient format. Each time a user asks for a web search using a search phrase, the search engine determines all the pages on the web that contain the words in the search phrase. However, organizing is only part of the problem. How do you determine what information is useful and what information is not? What is needed is a means of ranking the importance of the documents so that the information can be sorted with what is most important at the top of the list of returned results.

One way to do this is to have humans determine the ranking by looking for web pages that consist of a large number of links to other web pages in a particular area of interest. This list would have to be maintained constantly, as it may quickly fall out of date. Additionally, humans may unintentionally miss important pages. Another, and more efficient, way is to use Google's *PageRank* algorithm, a variant of eigenvector centrality aptly named after one of the algorithm creators, Sergey Brin and Lawrence Page (Brin and Page, 1998). Google uses network science to solve the problem without the use of humans to evaluate the web page content.

Web page production is a process by which an author creates a document that connects to other documents and whose page is the target for other web pages. When authors link to other web pages, they affirm the importance of the pages they link to. In essence, they pass their importance onto the linked pages. Therefore, the importance of a page is not only its own degree but also the number of pages that link to it. Now, let the web pages be the nodes of a network. The links will be the edges. Each node will be assigned an eigenvector variant centrality measure, *pr*, for PageRank. Here is how it is calculated.

Example 2.5

Given the network of web pages in Figure 2.10, find the PageRank of each node.

We begin by creating an adjacency matrix, $\mathbf{A} = [a_{ij}]$, of Figure 2.10

$$
\mathbf{A} = \begin{bmatrix}
0 & 1 & 1 & 0 & 0 & 0 & 0 & 0 \\
0 & 0 & 0 & 1 & 0 & 0 & 0 & 0 \\
0 & 1 & 0 & 0 & 1 & 0 & 0 & 0 \\
0 & 1 & 0 & 0 & 1 & 1 & 0 & 0 \\
0 & 0 & 0 & 0 & 0 & 1 & 1 & 1 \\
0 & 0 & 0 & 0 & 0 & 0 & 0 & 1 \\
1 & 0 & 0 & 0 & 1 & 0 & 0 & 1 \\
0 & 0 & 0 & 0 & 0 & 1 & 1 & 0
\end{bmatrix}
$$

The adjacency matrix, $\mathbf{A}$, gives the degree of each node, that is, how many connections it has to other nodes. PageRank is asking a slightly different question. It says how many connections a node has *from* other nodes, practically the reverse

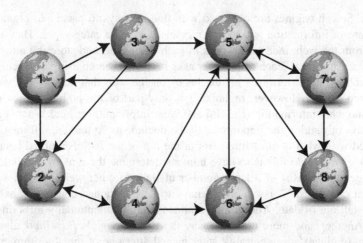

FIGURE 2.10 Network of eight hyperlinked web pages.

of the adjacency matrix. Mathematically, we reverse the adjacent matrix by taking its transpose, $\mathbf{A}^T = [a_{ji}]$. Notice the position of indices i and j.

$$\mathbf{A}^T = \begin{bmatrix} 0 & 0 & 0 & 0 & 0 & 0 & 1 & 0 \\ 1 & 0 & 1 & 1 & 0 & 0 & 0 & 0 \\ 1 & 0 & 0 & 0 & 0 & 0 & 0 & 0 \\ 0 & 1 & 0 & 0 & 0 & 0 & 0 & 0 \\ 0 & 0 & 1 & 1 & 0 & 0 & 1 & 0 \\ 0 & 0 & 0 & 1 & 1 & 0 & 0 & 1 \\ 0 & 0 & 0 & 0 & 1 & 0 & 0 & 1 \\ 0 & 0 & 0 & 0 & 1 & 1 & 1 & 0 \end{bmatrix}$$

Taking the transpose is important, but we also have to scale the columns so that we quantify the correct proportion of a node's total importance given to another node. This is done by dividing each element in the column by the corresponding node's out-degree. For instance, node 1 in Figure 2.10 has an out-degree of 2. Therefore, column 1 of $\mathbf{A}^T$ will be scaled by 2. Again, mathematically, we can capture the idea, and create a new matrix called **H**, which is the matrix of hyperlinks in a network of web pages defined as

$$\mathbf{H} = H_{ij} = \frac{a_{ji}}{\sum_{k=1}^{n} a_{kj}}$$

We will now create **H** for Figure 2.10.

$$\mathbf{H} = \begin{bmatrix} 0 & 0 & 0 & 0 & 0 & 0 & 0.33 & 0 \\ 0.50 & 0 & 0.50 & 0.33 & 0 & 0 & 0 & 0 \\ 0.50 & 0 & 0 & 0 & 0 & 0 & 0 & 0 \\ 0 & 1.00 & 0 & 0 & 0 & 0 & 0 & 0 \\ 0 & 0 & 0.50 & 0.33 & 0 & 0 & 0.33 & 0 \\ 0 & 0 & 0 & 0.33 & 0.33 & 0 & 0 & 0.50 \\ 0 & 0 & 0 & 0 & 0.33 & 0 & 0 & 0.50 \\ 0 & 0 & 0 & 0 & 0.33 & 1.00 & 0.33 & 0 \end{bmatrix}$$

Immediately we see by our construction of **H** that it has all non-negative values. Further, the sum of all entries in the column is 1 unless a page has no links. These two properties make **H** a special type of matrix called a *stochastic* matrix. If we define a vector, **p**, whose components are *pr*, PageRank can be found for every node in the network by solving the eigenvalue problem

$$\mathbf{p} = \frac{1}{\lambda}H\mathbf{p} \quad \text{or} \quad H\mathbf{p} = \lambda\mathbf{p}$$

One final property of **H** that is important to note is the fact that it is *stochastic*, which implies that its largest eigenvalue will always be 1. In other words, the vector, **p**, is an eigenvector of **H** with eigenvalue 1. This is also known as a *stationary* or *steady-state* vector of **H**. Using $\lambda = 1$, and appending a column of zeros gives the rank-deficient augmented matrix

$$\mathbf{H} - \mathbf{I} = \begin{bmatrix} -1 & 0 & 0 & 0 & 0 & 0 & 0.33 & 0 & 0 \\ 0.50 & -1 & 0.50 & 0.33 & 0 & 0 & 0 & 0 & 0 \\ 0.50 & 0 & -1 & 0 & 0 & 0 & 0 & 0 & 0 \\ 0 & 1.00 & 0 & -1 & 0 & 0 & 0 & 0 & 0 \\ 0 & 0 & 0.50 & 0.33 & -1 & 0 & 0.33 & 0 & 0 \\ 0 & 0 & 0 & 0.33 & 0.33 & -1 & 0 & 0.50 & 0 \\ 0 & 0 & 0 & 0 & 0.33 & 0 & -1 & 0.50 & 0 \\ 0 & 0 & 0 & 0 & 0.33 & 1.00 & 0.33 & -1 & 0 \end{bmatrix}$$

By elementary row operations as in the previous example, we find vector **p** to be

$$\mathbf{p} = \begin{bmatrix} pr\ 1 \\ pr\ 2 \\ pr\ 3 \\ pr\ 4 \\ pr\ 5 \\ pr\ 6 \\ pr\ 7 \\ pr\ 8 \end{bmatrix} = \begin{bmatrix} 0.2034 \\ 0.2288 \\ 0.1017 \\ 0.2288 \\ 0.3305 \\ 0.6864 \\ 0.6102 \\ 1.0000 \end{bmatrix}$$

This shows that page 8 is the most popular web page gaining the most influence from all its neighbors in the network (Austin, 2002).

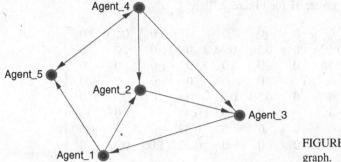

FIGURE 2.11 Directed graph.

2.3   DIRECTED NETWORKS AND CENTRALITY MEASURES

Directed networks are an important part of understanding centralities.

Degree centralities branch into an *in-degree* centrality and an *out-degree* centrality. To calculate the in-degree for node j, we simply add the number of links that terminate at j from other nodes, that is, that come *in* to node j. Similarly, to calculate the out-degree for node j, we add the number of links that originate at node j and terminate at other nodes.

Betweenness and *closeness* centralities rely on paths and geodesics for calculations. All that changes for these calculations is that when we consider paths within the network, we must travel in the direction of the links.

Eigenvector centrality is a more sophisticated version of degree centrality so it is certainly affected by directionality. Where degree centrality gives a simple count of the number of connections per node or agent in the network, eigenvector centrality acknowledges that not all connections are of equal value. Connections to agents who are themselves highly connected will give an agent more influence than connections to agents with few connections. Again, path is considered and travel must be in the direction of the link.

Example 2.6

Let us look at Figure 2.11 to see the difference in centrality measures between directed and undirected networks. After considering the directionality of the links, we can see that {5,4,2,3,1} and {3,1,2} are paths, whereas {5,4,5}, {5,1,3}, and {4,1,5} are not. The rest of the calculations are equivalent to undirected networks.

2.4   LOCATION IN THE NETWORK

Let us have a look at an example network of people who work together to see the importance of each of these measures. The following network is a network of researchers who are working on specific projects. The nodes are people and the links represent daily consultation to discuss work matters. If people do not meet

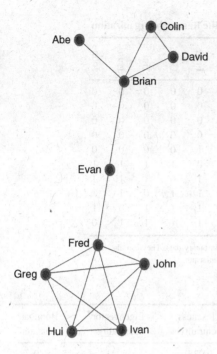

FIGURE 2.12   Network graph for the research organization.

to discuss work at least once per day, then there is no link between them. As people meet together, the relationship is reciprocated. That means that the links will be undirected. Figure 2.12 illustrates the network graph and Table 2.4 details the adjacency matrix for the relationships. Degree centrality indicates the people with the largest number of direct links to and from other people. Total degree is a combination of *in-degree* (links inwards from others) and *out-degree* (links outwards to others). Fred has the highest degree centrality, so he has a direct link to the highest number of people in the network. Fred has a total of 10 links to other people. We must remember to count both in and out links because the relationship are reciprocated. This means that Fred is central to the network and can directly influence the most people. Fred could be considered a "hub" as he has the largest number of people connecting to him. Abe has a total degree centrality of 2, which is the least number of connections as he only connects with Brian. In the real world, Abe could be working on a special task, or he could be isolated from the others like, say, a remotely located subcontractor. Brian, Greg, Hui, Ivan, and John, on the other hand, all have a total degree centrality of 8 because they all discuss matters with each other and Fred on a daily basis. They form a nice coherent group, but no one other than Fred has any connections to the others in the organization. Even Fred does not have a direct connection to Brian, Abe, Colin, and David. Although one might think that having a high degree gives dominance or power this is not always the case. A high total degree centrality does not necessarily mean significant power. It just indicates how strong the links are with adjacent nodes.

Betweenness centrality measures how often a given node is on the path between other nodes. It indicates the amount of control a node has over what

TABLE 2.4 Adjacency Matrix for the Research Organization

A	Abe	Brian	Colin	David	Evan	Fred	Greg	Hui	Ivan	John
Abe	0	1	0	0	0	0	0	0	0	0
Brian	1	0	1	1	1	0	0	0	0	0
Colin	0	1	0	1	0	0	0	0	0	0
David	0	1	1	0	0	0	0	0	0	0
Evan	0	1	1	0	0	1	0	0	0	0
Fred	0	0	0	0	1	0	1	1	1	1
Greg	0	0	0	0	0	1	0	1	1	1
Hui	0	0	0	0	0	1	1	0	1	1
Ivan	0	0	0	0	0	1	1	1	0	1
John	0	0	0	0	0	1	1	1	1	0

Adjacency matrices need not always contain binary data. They can also be weighted conveying the strength of a particular tie.

TABLE 2.5 Summary of the Centrality Measures

	Degree centrality	Betweenness centrality	Closeness centrality	Eigenvector centrality	Boundary spanner
Abe	2	0	0.038	0.021	0
Brian	8	40	0.056	0.085	1
Colin	4	0	0.040	0.028	0
David	4	0	0.040	0.028	0
Evan	4	40	0.063	0.267	1
Fred	10	40	0.063	1.000	1
Greg	8	0	0.048	0.947	0
Hui	8	0	0.048	0.947	0
Ivan	8	0	0.048	0.947	0
John	8	0	0.048	0.947	0

passes through to other parts of the network. Brian, Evan, and Fred are high in betweenness in this network as they are on the path that flows between the other people. Hence, they are brokers because they control what can flow (or not flow) through them to other parts of the network. Abe must go through Brian to contact anyone else in the network. The links between Colin and David are local and both must also go through Brian to contact the others in the network. Similarly, the links between Greg, John, Ivan, and Hui are also local and all must go through Fred to connect with other people in different parts of the network. Fred is well located with regard to betweenness as well as degree, but he is also a single point of failure. If he is absent or chooses to not pass on information, then the network could fragment. He may also choose to misrepresent information or pass on only selected parts of important information. Note that from Table 2.5, Fred, Brian, and Evan, all have betweenness scores of 40. Of the three, however, Fred can be considered the more significant of the three because its degree centrality is greater.

Closeness centrality measures how quickly a given node can reach all other nodes in the network and indicates how readily a node can access information and flows that are available from the network. Evan and Fred have the highest closeness measures, indicating they are the closest to all other people in the network. Hence, they can reach all the nodes in the network more efficiently than others. If you need to disperse information quickly across the network, then Evan and Fred can do this more effectively. Those with high closeness have the best view of what is happening in the network.

People in the network with high betweenness and high closeness are in a position where they have easy access to others in the network, while at the same time being able to control what is filtered through to other sections of the network. In our research network, we see that Evan and Fred have the highest closeness and betweenness scores, thus giving them some power over the others. Brian is also high in betweenness and closeness, and informally can exert some power.

Eigenvector centrality measures how well a node is connected to other well-connected nodes. In our network, Fred has an eigenvector of 1, the maximum obtainable. Fred is directly connected to Evan who is connected to Brian and the smaller research group, and he is connected to Greg, Ivan, Hui, and John who are all well connected within the larger research cluster. Greg, Hui, Ivan, and John are also high in eigenvector as they are connected to each other and Fred who are all well connected. The people least connected to well-connected nodes are Abe, Colin, and David as they have the lowest eigenvector scores.

Boundary spanners connect their group to others in the network and their links to adjacent nodes are termed bridges. In our network, the boundary spanners are Fred, Brian, and Evan. These three people are more central in the entire network as their connections reach beyond local connections within their own groups or clusters. Therefore, people can be boundary spanners because they are in a position to connect clusters to others in the network or have memberships in overlapping groups. The removal of boundary spanners can also fragment a network into disparate groups. On the other hand, because boundary spanners have access to information and ideas from different clusters, they are in the unique position to be able to combine this information to generate new ideas and innovations. In addition, boundary spanners are able to collate information from the sources to which they connect to build a bigger picture of a situation for informed decision making and planning of strategies. In many cases, the boundary spanner may be the only person who has access to these sources of information, thus giving them unique knowledge.

Let us look at the situation where Fred becomes aware that Brian and/or Evan are filtering information he gets from the smaller research cluster. In order to have reliable research results, Fred needs reliable and accurate information. He directly approaches David to gain more information. We now have a link between Fred and David and the network structure has changed slightly. No longer are the two clusters separated. The new network graph looks like Figure 2.13. The adjusted centrality measures are shown in Table 2.6. Brian and Evan have clearly lost influence in the network, as reflected in the decrease in their betweenness scores. Evan has lost his status as a boundary spanner. David's betweenness score

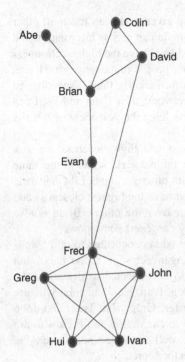

FIGURE 2.13 Network graph for the research organization.

TABLE 2.6 Comparisons of the Centrality Measures

	Degree centrality	Betweenness centrality	Closeness centrality	Eigenvector centrality	Boundary spanner
	Before/After	Before/After	Before/After	Before/After	Before/After
Abe	2/2	0/0	0.038/0.038	0.021/0.045	0/0
Brian	8/8	40/19	0.056/0.056	0.085/0.186	1/1
Colin	4/4	0/0	0.040/0.50	0.028/0.122	0/0
David	4/6	0/20	0.040/0.57	0.028/0.316	0/0
Evan	4/4	40/10	0.063/0.063	0.267/0.287	1/0
Fred	10/12	40/41	0.063/0.077	1.000/1.000	1/1
Greg	8/8	0/0	0.048/0.056	0.947/0.882	0/0
Hui	8/8	0/0	0.048/0.056	0.947/0.882	0/0
Ivan	8/8	0/0	0.048/0.056	0.947/0.882	0/0
John	8/8	0/0	0.048/0.056	0.947/0.882	0/0

This table shows the changes in the centrality measures for each of the people in this network.

has risen notably from 0 to 20 and Fred's from 40 to 41, making him the most influential in terms of betweenness. Colin and David are now higher in closeness, making it easier for them to directly contact others in the network. Fred is also higher in closeness as are all the members of Fred's cluster.

The removal of Evan or Brian in the old network structure would be devastating to the cohesion of the network, as removal of either would fragment the network into smaller unconnected groups and isolates. In the new network structure, the removal of Evan or Brian would not break the network apart. The new link between Fred and David has taken away the single point of failure observed in the old network structure. However, the removal of Fred from the network would still separate the network into two unconnected subnetworks.

Each additional relationship that Fred makes will require an investment of his time and energy. It may not be feasible for Fred to consult with every individual within the network, or his current relationships may consume all of his time. This will require Fred to be selective with regard to the connections he makes. In time, he may drop connections with Greg, Hui, Ivan, or John. These four individuals are *structurally equivalent*, meaning that they are connected to exactly the same alters. Thus, Fred's connections to these four individuals is likely to provide the same information and access to resources. In contrast, the connection to David provides access to a different part of the network. If Fred drops the connection to John, the network structure again changes to that shown in Figure 2.14. The resulting changes in centrality measures are shown in Table 2.7. The centrality scores for John are reduced because of the lost connection to Fred. Fred must now access the rest of the network through his connections to Greg, Hui, or Ivan. In turn, their centrality scores have increased. While Fred's degree centrality has become lower, as he is no longer connected to John, his betweenness centrality remains unchanged. Therefore, Fred is able to maintain power in the network in terms of his ability to control information and access resources, yet he has a lower cost in terms of time and effort to maintain consultation with John.

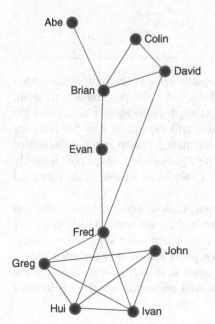

FIGURE 2.14 Network graph with the link between John and Fred removed.

TABLE 2.7   **Comparisons of the Centrality Measures**

	Degree centrality	Betweenness centrality	Closeness centrality	Eigenvector centrality	Boundary spanner
	Before/After	Before/After	Before/After	Before/After	Before/After
Abe	2/2	0/0	0.038/0.037	0.045/0.064	0/0
Brian	8/8	19/19	0.056/0.053	0.186/0.243	1/1
Colin	4/4	0/0	0.050/0.48	0.122/0.161	0/0
David	6/6	20/20	0.057/0.63	0.316/0.366	0/0
Evan	4/4	10/10	0.063/0.059	0.287/0.324	0/0
Fred	12/10	41/41	0.077/0.071	1.000/0.978	1/1
Greg	8/8	0/4	0.056/0.056	0.882/1.000	0/0
Hui	8/8	0/4	0.056/0.056	0.882/1.000	0/0
Ivan	8/8	0/4	0.056/0.056	0.882/1.000	0/0
John	8/6	0/0	0.056/0.042	0.882/0.795	0/0

This table shows the resulting changes in centrality measures when the link between Fred and John is removed.

While degree centrality may appear at first to look like an intuitive measure of influence, it can often be misleading. In the simple example provided, degree centrality represents a social cost of time and effort. Rather than influence, it is more a measure of constraint and loss of freedom of action. Betweenness centrality provides a more effective measure of power in the network.

So we can see that location in the network is very important for those people who either seek to be informed, collect information available via the network, or those who are presented with opportunities to control information flows across the network.

2.5   SUMMARY

Measures of centrality give us information regarding the position of agents within a network as well as the importance and influence that this position affords them. However, it must be borne in mind that the strength of an agent's location in the network as indicated by its centrality measures will depend on how the links are defined and measured. With any research, care must be taken in the definition of data and how it is measured. The famous saying "garbage in garbage out" is highly pertinent in network analysis. Your results will only be as reliable as the data used as the basis for the analysis.

Centrality measures of degree, betweenness, closeness, and eigenvector can tell us how much influence agents have within their network. Degree centrality indicates how many agents connect directly to the agent under study. Betweenness centrality reports how frequently an agent is on paths that are between other agents. Closeness centrality indicates how close an agent is to all the other agents in the network. Eigenvector centrality portrays how well connected our agent is to other well-connected agents.

Here is a summary of the key points discussed in this chapter:

- Degree centrality indicates the agents with the largest number of direct links to and from other agents. Agents with high degree are directly connected to more agents, either by incoming (in-degree) or outgoing (out-degree) information flows.

- Betweenness centrality measures how often a given node is on the path *between* other nodes and indicates the amount of control a node has over what passes through to other parts of the network.

- Closeness centrality measures how quickly a given node can reach all other nodes in the network and indicates how readily a node can access information and flows that are available from the network.

- Eigenvector centrality measures how well a node is connected to other well-connected nodes. Eigenvector emphasizes the saying, "It is not *what* you know but *who* you know."

- Boundary spanners connect *groups* to others in the network or have membership in overlapping groups. These agents are well positioned for control of information flow between groups and gathering information from different parts of the network. Spanners can be single points of failure in a network and their removal can result in fragmentation of the network into subnetworks.

- The position of agents in the network and their centrality will affect how much importance and influence they have (Brin and Page, 1998).

CHAPTER 2   LAB EXERCISE

Network Centrality Measures Using the Organizational Risk Analyzer (ORA) Software System

Learning Objectives

1. Create a new meta-network and enter network data in ORA.
2. Network visualization:
   (a) Size nodes by measure or attribute.
   (b) Color nodes by measure or attribute.
   (c) Size links by weight.
   (d) Represent directed networks and show labels.
3. Calculate centrality measures using the All Measures Reports in ORA.
4. Use charts in ORA.

1. Create a New Meta-Network

We are going to create the following five-node network (same as Figure 2.15).

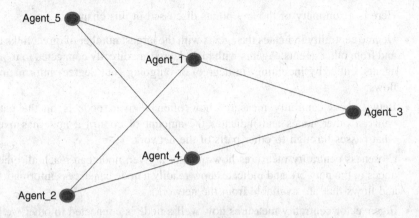

The network links are summarized in the following adjacency matrix:

	Agent_1	Agent_2	Agent_3	Agent_4	Agent_5
Agent_1	0	1	1	0	1
Agent_2	1	0	0	1	0
Agent_3	1	0	0	1	0
Agent_4	0	1	1	0	1
Agent_5	1	0	0	1	0

Open ORA and select the icon for a new meta-network.

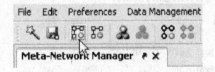

Select the icon for a new node class.

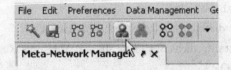

Leave the node type and ID as Agent. Enter 5 for the Size field in the Create Node Class dialog box to generate five nodes corresponding to the five nodes in the example problem. Then select *Create*.

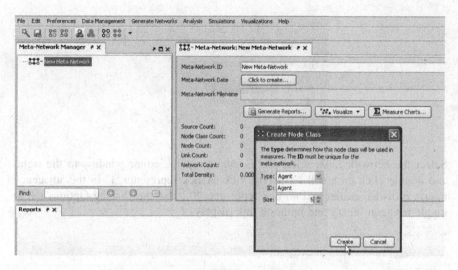

Next, select the icon for a new network.

Enter the name *Link* for the relationship in the dialog box. Select *Create.*

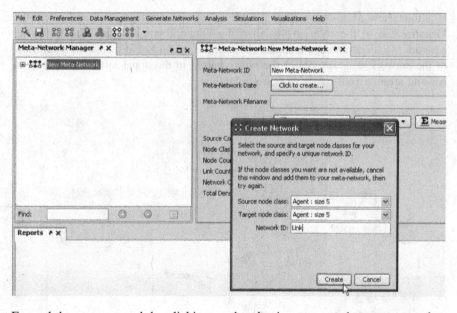

Expand the meta-network by clicking on the plus icon next to the meta-network.

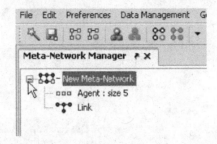

Select the network *Link*; select the *Editor* tab in the editor window to the right; and double click on a cell to insert a tick. A tick represents '1' in the adjacency matrix provided earlier and no tick represents '0.' Use the Display Options tab to toggle between binary and numeric link display.

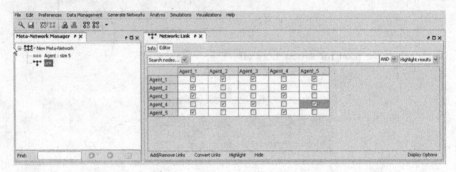

2. Network Visualization

Highlight New Meta-Network in the Meta-Network Manager area and select the *Visualize* tab to the right to visualize the network you entered. The following network visualization should appear. Move the Legend box if it restricts your view of the network diagram. Select the *pause* button in the upper left under the menu bar and arrange your nodes as you see below.

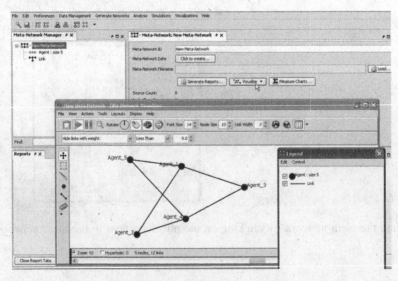

Attributes become important in many social network contexts. You can add attributes to the nodes in your network by using the node editor. Minimize the visualizer, select the *Agent* node class, select *Editor* and we are now ready to add attributes to Agent nodes. When you select the Editor tab you will see the Node ID and Node Title for the nodes. To the right are several options for manipulating nodes, attributes, and meta-networks.

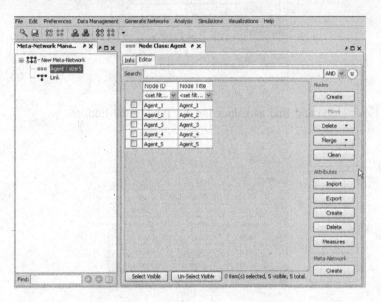

The first set of options allows you to create nodes and delete nodes. If you find aliases in your node data (i.e., two nodes are actually supposed to be the same person), then you can merge the nodes into a single node. The move option allows you to move a node into a different node class. If you move a node, you will be prompted on how to transfer links associated with this node.

The second set of options allows you to create, delete, or import attributes. Select *Create* under the *Attributes* utility. You are able to specify one of six types of attributes. Text and Number Categories will allow you to visualize these values in the visualizer, which we will do in a moment. The URI allows you to specify an internal website to list additional information about the node. The number will allow you to size the node based on the relative value of the numbers entered. For example, if you listed IQ as a number, you could size the node in the visualizer such that nodes with higher IQ appear larger than those with low IQ.

Select a Text Category, label it Type. Then select Create, and a new column will appear with the header Type. Select Close.

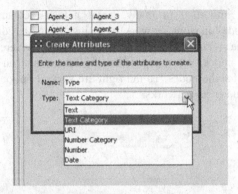

Now Enter Good and Bad as values for the Type in the Editor.

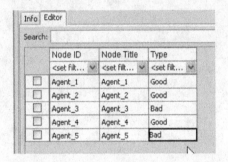

Next, *Visualize* your network as you did before. This time select *Display*, then *Node Appearance*, then *Node Color*, then *Color Nodes by Attribute* or *Measure*.

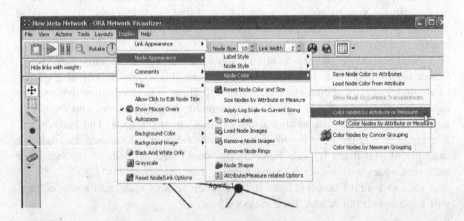

When the dialog box appears, select *Attribute* and then select the attribute name that you entered previously. You will see that the nodes become colored by the attribute. You can click on the color square and change the color of the node.

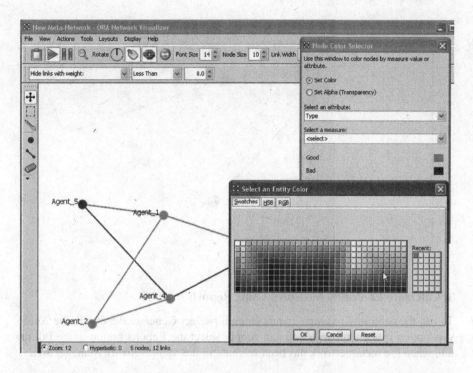

You can also size the node by a number attribute or by a centrality measure. Again, select *Display*, *Node Appearance*, then *Size Nodes by Attribute or Measure*. Select the measure or attribute that you wish to use for sizing criteria in order to size the nodes. Choose to size the node by *Centrality, Betweenness*. Agent_1 and Agent_4 are high in Betweenness centrality so these nodes are much larger than the others.

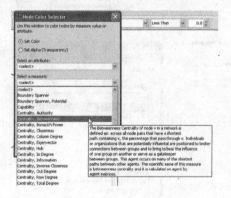

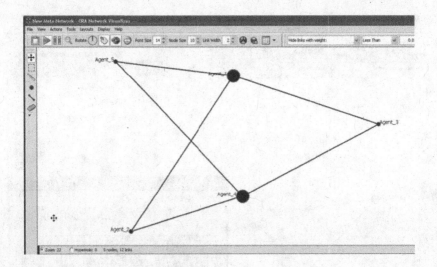

3. Calculate Centrality Measures Using Reports in ORA

To run the centrality measures, you can either select Generate Reports in the Analysis option in the toolbar or you can simply select the Reports button in the Editor tab, when the meta-network has been selected in the Meta-Network Manager area.

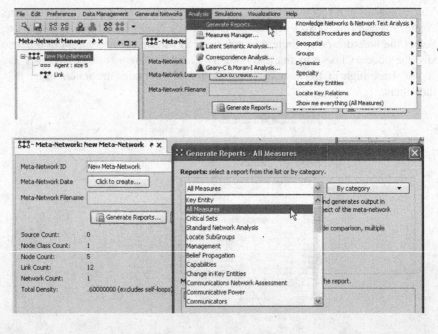

Select *All Measures* from the Select a Report pull-down menu and then click on the *Next* button. A dialog box will open that allows the analyst to set some options on how top central nodes will be displayed. Use the defaults and click on the *Next*

button. The final dialog box provides standard output options for all reports in ORA. The text button will produce a text file of the output. The CSV button produces a format that can be loaded into MS Excel. *HTML* is a user friendly format, so select this format for this exercise. Enter a filename for the report output and select *Finish*.

An output report will appear on your computer using your default Internet browser. This does not mean that you are necessarily connected to the Internet; it just allows the software to make use of some convenient display options.

Select the *Analysis for node class* Agent. You should see a display that appears something like the following:

Agent-level Measures

Input data: New Meta-Network

Start time: Thu Jan 05 11:44:02 2012

Return to table of contents

	Boundary Spanner: Link	Boundary Spanner, Potential: Link	Capability: Link	Centrality, Authority: Link	Centrality, Betweenness: Link	Centrality, Betweenness: Link, [unscaled]	Centrality, Bonacich Power: Link
Agent_1	0.000	0.3333*	0.9933*	1.000	0.2500*	3.0000*	3.0000*
Agent_2	0.000	0.111	0.841	1.000	0.056	0.667	2.000
Agent_3	0.000	0.111	0.841	1.000	0.056	0.667	2.000
Agent_4	0.000	0.3333*	0.9933*	1.000	0.2500*	3.0000*	3.0000*
Agent_5	0.000	0.111	0.841	1.000	0.056	0.667	2.000
MIN	0.000	0.111	0.841	1.000	0.056	0.667	2.000
MAX	0.000	0.333	0.993	1.000	0.250	3.000	3.000
AVG	0.000	0.200	0.902	1.000	0.133	1.600	2.400
STDDEV	0.000	0.109	0.075	0.000	0.095	1.143	0.490
GINI-COEFFICIENT	0.000	0.267	0.040	0.000	0.350	0.350	0.100

The measures are listed in alphabetic order, so you need to scroll down to find the specific measures you are seeking. You can see that ORA presents many more measures than those covered in this book and each measure is supported by academic literature on its proper use and interpretation. Covering the application of these measures is well beyond the scope of this book. The reader is referred to the ORA Users Guide and other publications available at www.casos.cs.cmu.edu or the ORA help menu for more information on other measures.

This report lists the measures for each agent in the network. Descriptive statistics on the minimum, maximum, average, and standard deviation are also provided for each measure. At this point, verify that the measures you previously calculated by hand agree with the measures calculated by ORA.

Now we will look at a larger network. The data for a much larger organization meta-network, the *Health Organization*, can be found in Appendix B, Tables B.3–B.16. Enter this data into ORA using the same steps as before. Save the meta-network as HealthOrganization.xml. You should have 43 agents and 2 networks. This meta-network represents a health organization where the nodes are health specialists and the two networks reflect who they go to for advice on clinical and administration matters. Visualize the entire meta-network. Note the number of nodes and links appears at the bottom of the visualization screen.

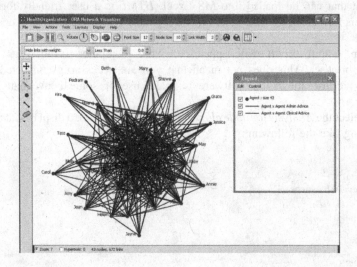

Highlight the Agent × Agent Clinical Advice network and visualize only this network.

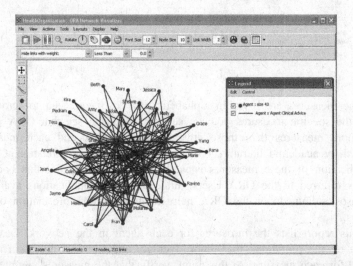

Then do the same for the Agent × Agent Admin network.

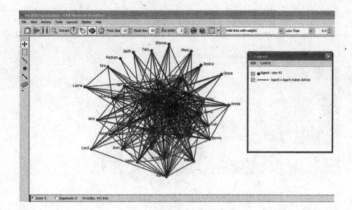

You can see that the network diagram is increasingly difficult to analyze visually as the number of nodes and links grows. However, we can get information about the network from different reports.

Let us run a couple of reports on the *HealthOrganization.xml* meta-network. We need to know who are the most important people in this network, so we will run the Key Entities report. Choose *Generate Reports* for the meta-network and select *Key Entities*. Make sure that only the HealthOrganization meta-network is selected for analysis. Continue through the input screens using the default values and produce the report in *HTML* format and select *Finish*. ORA will produce a report something like this:

KEY ENTITY REPORT

Input data: HealthOrganization

Start time: Thu Jan 05 19:58:22 2012

Table of Contents

Key Who - identifies the central actors and organizations

Performance Indicators - measures performance of the organization(s) as a whole

Produced by ORA developed at CASOS - Carnegie Mellon University

Select *Key Who identifies central actors and organizations* and the report should look like the image below.

The chart plots the 10 people who are repeatedly top ranked in the measures with the percentage of measures indicating the top 3 measures. ORA then lists a number of measures calculated including the centrality and other measures used in the calculations.

As you progress down the report, you will see an explanation for each measure listed, followed by the results for that particular measure. The input networks are also shown, so the user can see which networks have been used in each given calculation. For example, *Potentially Influential* is based on betweenness centrality and *Leader of Strong Clique* is based on eigenvector centrality.

You will find the All Measures report and the Key Entity report valuable for ascertaining centrality measures and also who is important in a network.

Agents

This chart shows the Agent that is repeatedly top-ranked in the measures listed below. The value shown is the percentage of measures for which the Agent was ranked in the top three.

Recurring Top Ranked Agent - HealthOrganization

Measure List

Emergent Leader (cognitive demand)

In-the-Know (total degree centrality)

Number of Cliques (clique count)

Leader of Strong Clique (eigenvector centrality)

Acts as a Hub (hub centrality)

Acts as an Authority (authority centrality)

Potentially Influential (betweenness centrality)

Connects Groups (high betweenness and low degree)

Group Awareness (shared situation awareness)

Emergent Leader (cognitive demand)

Measures the total amount of cognitive effort expended by each agent to do its tasks. need to move, connecting others, and so on. Such individuals may never become the formal leader of a group. Emergent leaders are identified in terms of the amount of cognitive effort that is inferred to be expended based on the individual's position in the meta-network. Individuals who are strong emergent leaders are likely to be not just connected to many people, organizations, tasks, events, areas of expertise, and resources; but also, are engaged in complex tasks where they may not have all the needed resources or knowledge and so have to coordinate with others, or have other reasons why they need to coordinate or share data or resources. The scientific name of this measure is cognitive demand and it is calculated on the agent by agent matrices.

Input network(s): Agent x Agent Admin Advice

Rank	Agent	Value
1	Rassy	0.977
2	Janet	0.744
3	Yang	0.744
4	Kaye	0.628
5	Kerry	0.605
6	Jani	0.488
7	Rana	0.442
8	Shar	0.372
9	Melanie	0.372
10	Angela	0.372

4. Use Charts in ORA

Network visualizations can be very powerful for you as an analyst, but often simple charts can be more convincing to general audiences. To generate charts of measures, click the *Measure Charts* tab and follow the prompts.

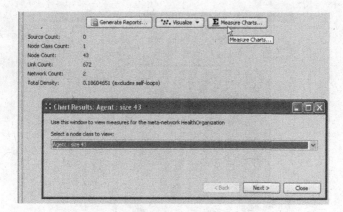

We recommend using bar charts, histograms, and scatter plots to illustrate your "points" (pun intended). A bar chart is a simple representation of each node you select and its corresponding measure. For example, if we wanted to display the top 5 degree centrality nodes, we can go to the bar chart and get a nice pictorial representation. The following figure presents the top 10 degree centrality nodes from the Agent × Agent Admin Advice network:

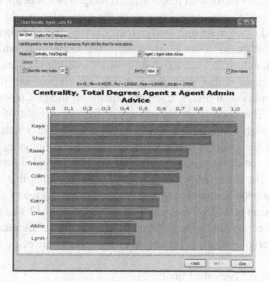

If we want to know what the distributions of degree centrality looks like, the histogram is a good feature to use. Clicking on the histogram tab gives us the following visualization:

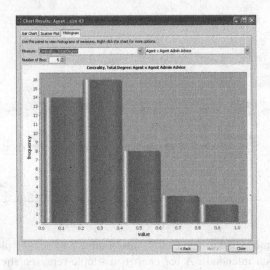

Let us suppose, however, that we want to see whether a nodes degree central-ity and betweenness centrality are highly correlated. If we were to look at a network of resource exchanges between and among an organization's staff, such a question would tell us whether those who frequently exchange resources are the same people who exchange resources between clusters in the network (i.e., do frequent traders trade locally or do they span and broker trades across the organization).

To study this question, we would look to the scatter plot, which simply plots each of these quantities on each axis and tells us whether they are correlated.

In the example scatter plot, the x-axis is the degree and the y-axis is the betweenness of individual nodes. Each red square on the plot represents the nodes in the network. The nodes that are plotted in the upper right of the scatter plot show unusually high betweenness. These nodes are the individuals in the example that span across the organization and serve as brokers. Statistics to describe the relationship between degree and betweenness are also provided just above the plot. In this example, the slope (M) of the line is 2.1888, meaning that the standardized betweenness of a node is roughly twice the standardized degree of a node. The r^2 of the linear regression is 0.1187, meaning that approximately 12% of the betweenness of a node can be explained by the degree of the node. There are quantities indicated on this graph that give us clues about the relationships between the two quantities we are studying.

The results are presented here:

Symbol	Meaning
r^2	What percentage of variation in Y is explained by X?
m	A one unit increase in X means a m unit increase in Y.
b	If $X = 0$ then $Y = b$.

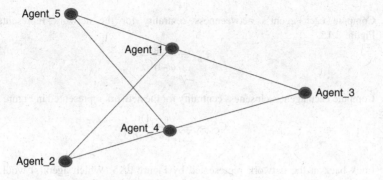

FIGURE 2.15   Centrality example graph.

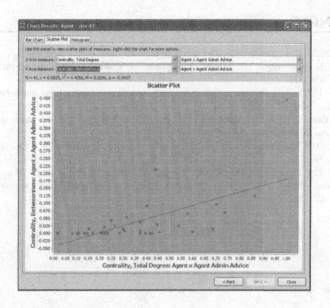

EXERCISES

2.1. Create the adjacency matrix for the network represented in Figure 2.15.

2.2. Compute the degree centralities for each agent in the network represented in Figure 2.15.

$$C_{Di}^* = \frac{\sum_{j=1}^{n} a_{ij}}{(n-1)}$$

2.3. Calculate total possible geodesics for each agent in the network represented in Figure 2.15.

2.4. Compute each agent's betweenness centrality for the network represented in Figure 2.15.

$$C_{Bi}^* = \frac{C_{Bi}}{_{(n-1)}C_2}$$

2.5. Compute each agent's closeness centrality for the network represented in Figure 2.15.

$$C_{Ci}^* = \frac{(n-1)}{\sum_{j=1}^{n} d(n_i, n_j)}$$

2.6. Look back at the network represented by Figure 2.15. Which agent(s) would you develop for further inquiry? Why?

REFERENCES

Austin, D. (2002). How google finds your needle in the web's haystack. Found online.

Barabási, A.-L. and Albert, R. (1999). Emergence of scaling in random networks. *Science*, **286**(5439):509–512.

Bonacich, P. (1972). Factoring and weighing approaches to clique identification. *Journal of Mathematical Sociology*, **2**(1):113–120.

Brin, S. and Page, L. (1998). The anatomy of a large-scale hypertextual web search engine. *Computer Networks and ISDN Systems*, **30**(1–7):107–117.

Freeman, L. (1979). Centrality in social networks conceptual clarification. *Social Networks*, **1**(3):215–239.

Frobenius, G. (1912). Ueber matrizen aus nicht negativen elementen. *Preuss. Akad. Wiss.*, 456–477.

Watts, D. and Strogatz, S. (1998). Collective dynamics of 'small-world' networks. *Nature*, **393**(6684):440–442.

GRAPH LEVEL MEASURES

We must all hang together, or assuredly we shall all hang separately.
—*Benjamin Franklin*

Learning Objectives

1. Understand graph level measures of density and diameter.
2. Understand centralization measures of degree centralization, betweenness centralization, and closeness centralization.
3. Understand average centrality measures for graphs of average degree, average betweenness, average closeness, and average eigenvector.
4. Explain the difference between a graph level measure and a node/agent level measure.
5. Calculate and interpret average centralities using ORA.

Why Graph Level Measures?

Rather than focus on agents within a social network, we might be interested in some of the properties of the network as a whole. How does the organization seem to get along collectively? What is the average path length for a message to travel from one agent to the next? Do agents in this network tend to make relational connections when introduced? Can we neatly separate the network into two teams? Are there key agents who are highly central to the network while the majority of the remaining agents are not high in centrality? What does the structure of the network tell us about the ability of this network to withstand the removal of highly central agents? How does this network compare to another network of the same size and type? These are questions that could perhaps be answered given a list of agent measures, but it is much easier to look at a few graph level measures, also referred to as network level measures, to find the answers to all these questions.

Social Network Analysis: with Applications, First Edition.
Ian A. McCulloh, Helen L. Armstrong, and Anthony N. Johnson.
© 2013 John Wiley & Sons, Inc. Published 2013 by John Wiley & Sons, Inc.

3.1 DENSITY

The density of a network is a graph or network level measure of the ratio of the number of links present given the total number of links possible. In a dense network, there are many links. In a sparse (opposite of dense) network, there are relatively few links. Before we determine how to calculate the density of a specific network, we must explore how to determine the total number of possible links that could be present in the network.

There are a couple of ways to understand how to determine the total number of possible links in a network. We have already explored this in the previous chapter when we were investigating betweenness centrality. If there are n nodes, then the number of potential combination of pairs of nodes is $_nC_2$, pronounced n choose 2. This expression can be further expressed as $_nC_2 = \frac{1}{2}n(n-1)$, and it describes the number of undirected or bi-directed potential links in the network. If the network is directed, then we must use a permutation instead of a combination. The potential pairs of nodes in a directed network can be expressed as $_nP_2 = n(n-1)$.

Another way to visualize the number of possible links is to consider an $n \times n$ adjacency matrix. To determine the total number of entries possible in a matrix, simply multiply $n \times n$, which is n^2. Recall that in an adjacency matrix, the reflexive (self links) are undefined and do not exist. Therefore, we must remove n possible entries from the n^2 options, which is $n^2 - n = n(n-1)$. This is equivalent to the number of permutations. If a link from node i to node j is the same as a link from node j to node i, then there are only half as many possible entries to consider. The equation becomes $\frac{1}{2}n(n-1)$, which is equivalent to the number of combinations. The density is then the number of links present in the network, divided by the total number of possible links.

Example 3.1

Consider a network with seven nodes. Figure 3.1 is a chart of the possible entries for links. There are seven nodes. Therefore, there exist 49 squares that could contain

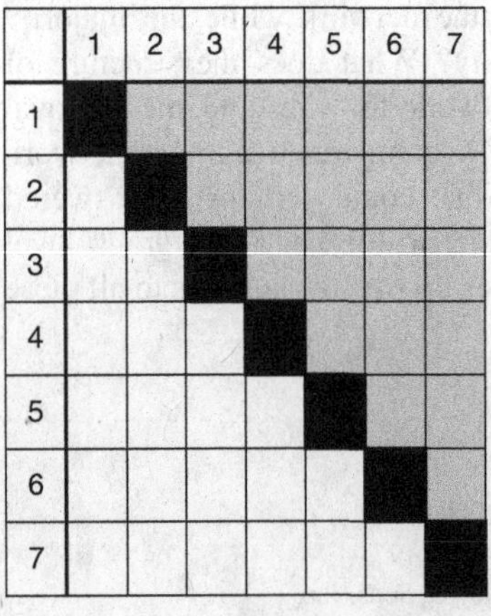

FIGURE 3.1 Network density chart.

an entry describing a relationship. The black squares represent the reflexive links that are typically excluded. This leaves $n(n-1) = 7(7-1) = 42$ squares. Half of the 42 squares are gray and the other 21 squares are white. If the network is undirected, then we would only consider the gray squares as these will be a mirror image of the white squares. If, however, the network is directed, then we must consider both the white and gray squares.

Example 3.2

Find the density of an undirected network consisting of 30 nodes and 145 links.

$$\text{Density} = \frac{L}{\frac{1}{2}n(n-1)} = \frac{145}{\frac{1}{2}(30)(30-1)} = \frac{145}{435} = \frac{1}{3}$$

Example 3.3

Find the density of a directed network consisting of 24 nodes and 46 links.

$$\text{Density} = \frac{L}{n(n-1)} = \frac{46}{(24)(24-1)} = \frac{46}{552} = \frac{1}{24}$$

Check on Learning

If there are 10 links in a seven-node undirected network, calculate the density.

Answer

$$\frac{L}{\frac{1}{2}n(n-1)} = \frac{10}{\frac{1}{2}(7)(7-1)} = \frac{10}{21}$$

3.2   DIAMETER

The diameter is a network level measure of network connectedness. The diameter describes the length of time or amount of effort needed for information to move from one end of the network to the other. The diameter is simply the longest geodesic in the network. Recall that the geodesic is the shortest path between two nodes. The shortest geodesic will, of course, be equal to 1 in the case of adjacent nodes.

Example 3.4

Find the diameter of the network shown in Figure 3.2.

To find the diameter, simply look for the longest geodesic. Start with a node on the periphery of the network. For example, there is a geodesic of length one starting with agent 3, {3,4}. There is a geodesic of length 2 starting with agent

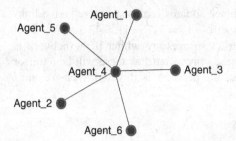

FIGURE 3.2   Star graph.

3, {3,4,5}. There are no geodesics of length 3 or more. Therefore, the maximum geodesic is length 2; thus, the diameter = 2.

Example 3.5

Find the diameter of the network shown in Figure 3.3.

To find the diameter, we again look for the longest geodesic. Start with a node on the periphery of the network. There is a geodesic of length 1 starting with agent 3, {3,4}. There is a geodesic of length 2 starting with agent 3, {3,4,5}. In this network, there is also a geodesic of length 3 starting with agent 3, {3,4,5,6}. There is no geodesic of length 4. The path {3,4,5,6,1} is not a geodesic because the path from agent 3 to agent 1 could also be reached through {3,2,1}, which is shorter than {3,4,5,6,1}. Thus, the diameter = 3.

Notice the difference in the two examples. The first example (the star graph) had a diameter of 2. In this network, agent 4 quickly passed information to all the other agents in one step. No agent in the network is farther away than two steps from any other agent. In the second example (the circle graph), however, there are no brokering agents. It would take longer for an agent to pass information to the rest of the network.

Check on Learning

What is the diameter of a circle social network consisting of six agents?

Answer

The longest geodesic will be length 2, so the diameter of the network is 2.

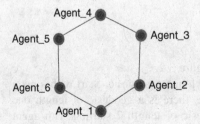

FIGURE 3.3   Circle graph.

3.3   CENTRALIZATION

Centralization provides a network level measure of potentially exceptional nodes in the network. In other words, is there a node in the network that is much more central than typical nodes? Centralization can be calculated for any of the centrality measures. Thus, there is degree centralization, betweenness centralization, and closeness centralization. Each centralization determines the degree to which the maximum individual nodal centralities exceed the centralities of the other nodes in the network. The index will range from 0 (no node exceeds the others) to 1 (one node exceeds all others). For expository purposes, we will calculate centralization measures for the two extremes, primarily the examples from the previous sections, labeled network A (star) and network B (ring) (see Fig. 3.4). Note that the "star graph" and "circle graph" are networks that highlight the extreme boundaries of all networks, and they are bidirectional (undirectional). Directional networks will have calculated values between 0 and 1 also. The general form of centralization given by Freeman (1979) is

$$NC_X = \frac{\sum_{i=1}^{n}(C_{X\,\max} - C_{Xi})}{\max \sum_{i=1}^{n}(C_{X\,\max} - C_{Xi})}$$

where n is the number of nodes, C_{Xi} represents the individual nodal centrality values, $C_{X\,\max}$ is the largest value of C_{Xi} for any node in the network, and $\max \sum_{i=1}^{n}(C_{X\,\max} - C_{Xi})$ equals the maximum possible sum of differences in nodal centrality for a network of n nodes.

3.3.1   Degree Centralization

Degree centralization reflects the relative dominance of a single node over all other nodes in the network. Networks with a high degree centralization measure are subject to fragmentation if these highly connected nodes are removed. Networks with a low degree centralization measure are less sensitive to fragmentation by the removal of nodes because no one node connects all other nodes in the networks.

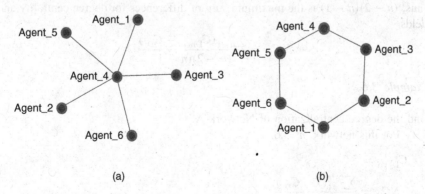

(a)   (b)

FIGURE 3.4   Two different networks: (a) network A and (b) network B.

The degree centralization formula as expressed by Freeman (1979) is given as

$$NC_D = \frac{\sum_{i=1}^{n}(C_{Dmax} - C_{Di})}{\max \sum_{i=1}^{n}(C_{Dmax} - C_{Di})}$$

The numerator of NC_D is calculated using the *nonstandardized* measure C_{Di} rather than the standardized (normalized) measure C_{Di}^*. The denominator is a calculation of the maximum sum of differences and is calculated by first noting that the maximum value of C_{Di}, that is, C_{Dmax} is $(n-1)$, which represents a network that contains one node lying adjacent to all of its neighbors. The neighbors of this one special node in turn would have only one link into the network. This is the exact representation of the star network of Figure 3.4a Network A. The structure of the graph dictates that all but one of the C_{Di} value are 1. The differences then become

$$(C_{Dmax} - C_{Di}) = (n - 1) - 1 = (n - 2).$$

Summing the differences will yield

$$\sum_{i=1}^{n}(C_{Dmax} - C_{Di}) = (n - 2) + (n - 2) + \cdots = (n - 2)(n - 1).$$

The question one might ask is whether or not this, in fact, is the maximum difference? To answer this question, we can suppose that we add a line to any pair of nodes of the star so that it becomes, say, a part of a rim of a wheel. This would make the connected nodes compared to the center yield a difference

$$(C_{Dmax} - C_{Di}) = (n - 1) - 2 = (n - 3)$$

which is smaller than $(n - 2)$. If we remove a link, then one node will be isolate, that is, entirely unconnected. The isolate node will yield a difference of

$$(C_{Dmax} - C_{Di}) = (n - 2) - 0 = (n - 2)$$

but all the other node differences will be reduced:

$$(C_{Dmax} - C_{Di}) = (n - 2) - 1 = (n - 3).$$

Thus, $(n - 2)(n - 1)$ is the maximum sum of differences for degree centrality and yields

$$NC_D = \frac{\sum_{i=1}^{n}(C_{Dmax} - C_{Di})}{(n - 2)(n - 1)}$$

Example 3.6

Find the degree centralization of Network A.
    For this network, $n = 6$.

$$C_{Dmax} = 5$$

$$NC_D = \frac{\sum_{i=1}^{6}(5 - C_{Di})}{(4)(5)}$$

$$= \frac{(5-1)+(5-1)+(5-1)+(5-5)+(5-1)+(5-1)}{20}$$

$$= \frac{20}{20} = 1$$

A centralization of 1 indicates that there is a single point of failure for the measure centrality in the network. In other words, one node is completely central, while other nodes are not central at all.

Example 3.7

Find the degree centralization of Network B.
For this network, $n = 6$.

$$C_{Dmax} = 2$$

$$NC_D = \frac{\sum_{i=1}^{6}(2 - C_{Di})}{(4)(5)}$$

$$= \frac{(2-2)+(2-2)+(2-2)+(2-2)+(2-2)+(2-2)}{20}$$

$$= \frac{0}{20} = 0$$

A centralization of 0 indicates that all the nodes have the same centrality and there is no node that is more influential than any other node.

3.3.2 Betweenness Centralization

Betweenness centralization will indicate whether a sole gatekeeper exists within the network, or whether no one node controls access to all the other nodes in the network. Betweenness centralization is calculated using the standardized indices (between 0 and 1) for betweenness and is defined as

$$NC_B = \frac{\sum_{i=1}^{n}(C_{Bmax}^* - C_{Bi}^*)}{(n-1)}$$

with the sum of maximum differences being equal to $(n-1)$.

Example 3.8

Find the betweenness centrality of Network A.

The betweenness centrality of Agent_4 is 1 and the betweenness centrality of all other nodes is 0. Therefore, $C_{Bmax}^* = 1$. Thus, the expression for betweenness centralization can be expressed as

$$\frac{\sum_{i=1}^{6}(1 - C_{Bi}^*)}{(6-1)} = \frac{1+1+1+0+1+1}{5} = \frac{5}{5} = 1$$

A centralization of 1 indicates that there is a single "gatekeeper" in the network. In other words, one node completely controls access to the other nodes. The other nodes do not control access to any other nodes.

Example 3.9

Find the betweenness centralization of Network B.

The betweenness centrality of all nodes in the network is the same and the value is 0.2. Therefore, $C_{\text{Bmax}}^* = 0.2$. Thus, the expression for betweenness centralization can be expressed as

$$\frac{\sum_{i=1}^{6}(0.2 - C_{Bi}^*)}{(6-1)} = \frac{0+0+0+0+0+0}{5} = \frac{0}{5} = 0$$

A centralization of 0 indicates that there is no single "gatekeeper" in the network. No one node controls access to the other nodes any more or less than any other node in the network.

3.3.3 Closeness Centralization

Closeness centralization will indicate the similarity of the closeness of nodes and will show whether there is a dominant node that is as close as only one step away from all other nodes in the network. Closeness centralization is calculated using the standardized indices for closeness and is defined as

$$NC_C = \frac{\sum_{i=1}^{n}(C_{C\max}^* - C_{Ci}^*)}{[(n-2)(n-1)/(2n-3)]}$$

with the sum of maximum differences being equal to $[(n-2)(n-1)/(2n-3)]$.

Example 3.10

Find the closeness centralization of Network A.

The closeness centrality of Agent_4 is 1 and the closeness centrality of all other nodes is 5/9. Therefore $C_{C\max}^* = 1$. Thus, the expression for closeness centralization can be expressed as

$$\frac{\sum_{i=1}^{6}\left(\frac{5}{9} - C_{Ci}^*\right)}{[(6-2)(6-1)/(2(6)-3)]} = \frac{4/9+4/9+4/9+0+4/9+4/9}{20/9} = \frac{20/9}{20/9} = 1$$

A centralization of 1 indicates that there is a single node that is one step away from every other node in the network. In other words, one node can pass information to every other node in a single step. The other nodes are at least two steps from each other.

Example 3.11

Find the closeness centralization of Network B.

The closeness centrality of all nodes in the network are the same and the value is 5/9. Therefore, $C_{max} = 5/9$. Thus, the expression for closeness centralization can be expressed as

$$\frac{\sum_{i=1}^{6} \left(\frac{5}{9} - C_{Ci}^{*} \right)}{[(6-2)(6-1)/(2(6)-3)]} = \frac{0+0+0+0+0+0}{20/9} = \frac{0}{20/9} = 0$$

A centralization of 0 indicates that all nodes are equidistant and, therefore, require the same number of steps to traverse the network. In other words, every node passes information to every other node in the same number of steps.

3.4   AVERAGE CENTRALITIES

The *averages* of some centralities at a network level can provide us with valuable information when comparing different networks. We will now look at average centralities for degree, closeness, betweenness, and eigenvector and see what these average measures mean.

Average degree centrality is a measure of network density

$$d = \frac{\sum C_{Di}}{n} = \frac{\sum (\sum_{j=1}^{n} a_{ij})/(n-1)}{n} = \frac{\sum (\sum_{j=1}^{n} a_{ij})}{n(n-1)}$$

which is a measure of the number of links present compared to the total number of links possible. Because of the way we have defined it here, this number will vary between 0 and 1 and is standardized to measure networks with varying sizes of n. Density is probably the most widely used network level measure and is regarded as a measure of group cohesion.

Average closeness centrality measures the average length of geodesics within the network.

$$\frac{\sum C_{Ci}}{n}$$

Average betweenness centrality is the average number of agents per geodesic.

$$\frac{\sum C_{Bi}}{n}$$

The average eigenvector centrality has a complicated mathematical expression, but it can be interpreted as follows: high average eigenvector centralities mean that, on average, agents are connected to other influential agents, that is, lower values of average eigenvector centrality indicate that there may be a small "elite group" in the network with much influence.

Check on Learning

1. What is the difference between average degree centrality and degree centralization?

Answer

1. Average degree centrality is a measure of the number of links present compared to the total number of links possible in the network. Degree centralization indicates the relative dominance of a single node over all other nodes in the network.

3.5   NETWORK TOPOLOGY

Network topology, which is the structure of the network as a whole, can tell us much about the characteristics of the network and how it might be expected to behave. Different network topologies can reflect different graph measures. The measures can be characterized by values such as their clustering coefficient, diameter, average shortest path, degree, degree distribution, and robustness. Airoldi and Carley (2005) summarizes the pure types of network topologies as follows:

- Ring lattice, where each node is connected to its neighbors.
- Small world, where each node is connected to some of its neighbors and a few distant nodes.
- Random (Erdös–Rényi), where each node is connected to a random set of the remaining nodes.
- Core-periphery, where nodes belong exclusively to either the core or the periphery, and the core and periphery nodes are connected to core nodes; however, there are no edges among periphery nodes.
- Scale free, where most of the nodes are connected to a few other nodes, while few nodes are connected to many other nodes, and formally described with a power law.
- Cellular, where nodes are divided into cells and connections are frequent between nodes within each cell, and rare between nodes in different cells.

Table 3.1 lists a summary of features for each topology.

3.5.1   Lattice Networks

The lattice topology is a completely ordered cubic structure predominantly reflected by electrical networks, swarms of insects, schools of fish, and flocks of birds where the behavior of each individual depends on the behavior of its near neighbors.

TABLE 3.1 Comparison of Network Topologies

	Lattice (Regular)	Small world	Core periphery	Cellular	Scale free	Random
Clustering coefficient	High and constant	High	Medium/low	Medium/Low	Medium/Low	Low
Distance (average shortest path, geodesic)	High	Low/medium	Medium	Medium/high	Medium	Medium
Diameter	Very High	High/Medium	Medium/Low	Medium/High	Low	Low/Medium
Degree (hubs)	Low	Medium	Variable	Variable	High—many with low degree, few with high degree	Low
Degree sequence distribution	Delta like	Poisson like	Power law like	Power law like	Power law distribution	Poisson
Robustness (removal of nodes)	High	High	Medium (if removal of non-hub nodes) Low (if removal of hubs)	Medium (if removal of non-hub nodes) Low (if removal of hubs)	Medium (if removal of non-hub nodes) Low (if removal of hubs)	High
Randomness (entropy)	Low	Low	Medium	Medium	Medium	High
Diffusion	Medium	High/ medium	Medium	Medium	High	Medium/ high

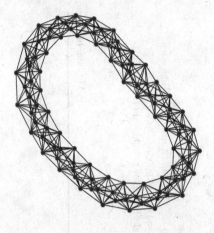

FIGURE 3.5   The lattice network topology.

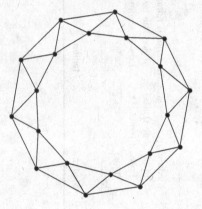

FIGURE 3.6   The mesh pattern in a lattice network.

Figure 3.5 shows the structure of a network with the lattice topology comprising 100 nodes. Owing to its highly structured geometric formation, the lattice architecture is seldom realized in the real world. In real-world networks, Watts, (2003) posits that real networks more often fall between the lattice and the random structures at the two extremes of the structural spectrum. The grid or mesh structure forming the foundation of the lattice can be seen in Figure 3.6; this network has 20 nodes, each with four connections. The geometric arrangement of the nodes can be clearly seen in this diagram. Notice the absence of links that connect nodes across the graph as is usually found in small world networks. This means that information must travel around the network until it reaches its destination, making the shortest path between nodes longer than that of a small world network. Lattice networks display high clustering, low degree, and long average path lengths as there are no shortcuts. It comprises many short steps and requires many hops to reach the outer edges of the network. The short links also contribute to the higher clustering and formation of local neighborhoods. Each node in a lattice network has the same centrality measures, due to the geometrically regular structure. Each node in the network in Figure 3.6 will exhibit the same centrality measurements (scaled to between 0 and 1): degree = 0.211, closeness = 0.345, betweenness = 0.105, and eigenvector =

1.000. The lattice topology is the most highly structured of topologies and the least random, situated at the opposite extreme to random (Erdös–Rényi) networks.

3.5.2 Small World Networks

Networks with the small world topology are characterized by nodes than can reach, and can be reached by, a small number of other nodes in a small number of steps. Small world networks are sometimes referred to as *six degrees of separation*. There is even a web site where an individual can determine the degrees of separation of actors from Kevin Bacon (http://www.thekevinbacongame.com/). A prominent example of a small world network is the Internet, where user nodes are connected to other nodes through hyperlinks, and users can make their way through the internet network via short steps from one site to another. Figure 3.7 illustrates the structure of a small world network with 100 nodes.

In mathematical terms, the distance between two nodes chosen at random in a small world network increases in proportion to the logarithm of the number of nodes in the network. Characteristically, small world networks have small geodesics (shortest distance between nodes) and large clustering coefficients, reflecting social networks that have short distances between people and high clustering with lots of cliques. Notice the connections of nodes across the graph in Figure 3.7. These links provide greater connectivity and shortcuts for the spread of information.

The Milgram experiment by Stanley Milgram (Travers and Milgram, 1969) was the most prominent research in this area of network structure. Milgram chose 296 individuals at random from the cities of Omaha, Nebraska and Wichita, Kansas and asked them to forward a letter he provided to a target person in Boston. The individuals chosen were asked to forward the letter to the target person in Boston if they knew them directly, or, alternatively, to forward it to a person they knew who was most likely to know this target person. Of the 20% of letters that successfully reached the target person in Boston, the average length of the path was 6.5, meaning it took 6.5 steps to move between connections across the network from the original

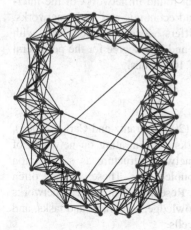

FIGURE 3.7 Small world network topology.

individual at the source city to the target in Boston. A later experiment by Dodds et al. (2003) used emails instead of letters, with 18 targeted individuals across 13 countries and 60,000 participants, resulting in an average path length of 4.

Substantive work has been undertaken in small world networks by Duncan Watts and Steve Strogatz (1998) and Kleinberg (2000, 2001). While small world networks are high in clustering and low in distance between nodes, they also indicate a high number of hubs (see Table 3.1) that could be susceptible to attack. However, the high clustering means that alternate connections can be made easily, increasing the shortest path by only a small amount. This ensures that small world networks are robust to attack.

Latora and Marchiori (2001) confirmed in their research that neural, communication, and transport networks illustrate small world characteristics. Other examples of networks illustrating small world characteristics include the brain, the *Caenorhabditis elegans* worm, airline flight paths, power grids, collaboration of film actors, and signal propagation.

3.5.3   Core Periphery

Core periphery network topologies are intermediate scale (mesoscale) architectures and comprise two sets of nodes: a core set where nodes are closely tied to one another, and a peripheral set who connect more often to core members rather than other peripheral members. Research conducted by Borgatti and Everett (2000) led the way in core-periphery network structure. They added a third set of nodes possible in the core periphery structure; the nodes in the network that do not fall within the ingroup (core) and outgroup (periphery).

The core commonly displays characteristics of a cohesive subgroup with close connections and clique formations. The peripheral nodes are only sparsely connected. The core nodes are also well connected to peripheral nodes, while nodes within the core are well connected and share information and events. Those in the periphery commonly connect only with core nodes and have no interconnections at the periphery layer. Core members thus have a structural advantage over the peripheral nodes. The core periphery structure has been found in networks of the international money market, geographical systems, and economic and social networks.

The density of core periphery networks differs between the node sets, with high density and degree in the set of core nodes, and low degree for the peripheral nodes. Figure 3.8 is a core-periphery network of 100 nodes.

3.5.4   Cellular Networks

Substantial work has been contributed by Krebs (2002), Frantz and Carley (2005), Frantz and colleagues (2009) to cellular networks concentrating on networks of terrorists. Frantz and Carley describe cellular network structure as a collection of distributed, but sparsely connected, tightly coupled cells. These cells are often small and operate independently of one another. Researchers in terrorist networks commonly find that cells specialize in certain knowledge, resources, and tasks, and often do not know about the activities of other cells.

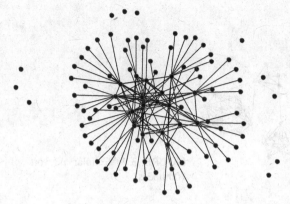

FIGURE 3.8   Core-periphery network of 100 nodes.

Krebs reported that covert networks are different from normal social networks in that terrorists minimize their connections within the network and do not forge new ties with those outside the network. The cells are tied by strong connections, often made during training; however, these ties frequently remain dormant and hidden and are only activated when needed. As such they appear to be weak ties in Granovetter's terms (Granovetter, 1973). These hidden links make the network structure difficult to discover and give it strength and resilience. The connections in covert networks take the form of trust relationships, tasks including phone calls, email, travel records, web sites, observations, and money and resources including bank account and money transfers.

The structure of cellular network topology results in a wide network diameter owing to the low number of links between cells. While connections within the cells can be frequent, the isolation of these cells from the rest of the network affects its average shortest path and overall cohesion. If key nodes in the cells can be identified and removed, then the cell often falls into disarray owing to the lack of connections to other cells and mastermind terrorists. Carley et al. (2001) details approaches to destabilizing terrorist networks by simulating the targeting of particular nodes and evaluating the results of their removal from the network. Figure 3.9 is a cellular network of 100 nodes.

3.5.5   Scale-Free Networks

Scale-free networks are characterized by a large number of nodes with low degree and a small number of nodes with high degree (hubs). In the scale-free architecture, new nodes to the network connect to other nodes by preferential attachment, not randomly. Hence, we find that scale-free networks have clustering around hub nodes, as new nodes will prefer to link with other well-connected nodes. Some examples of scale-free networks are the world wide web, protein networks, epidemiology, and ecosystems. For example, ecosystem elements are known to form preferential connections on the basis of energy efficiency, and connections to giant hubs such as Google and Amazon on the world wide web are preferred to connecting to relatively unknown web sites. Barabási and Albert (1999) coined the

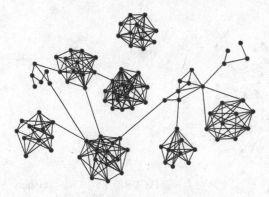

FIGURE 3.9 Cellular network of 100 nodes.

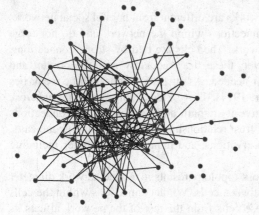

FIGURE 3.10 Scale-free network of 100 nodes.

actual phrase "scale-free network" as they researched the concept of preferential attachment on the world wide web and other networks with power law distributions.

Figure 3.10 illustrates the structure of a scale-free network with 100 nodes. Notice the high number of isolate nodes and those with only one or two links to other nodes. There are a few nodes toward the center of the graph that have many connections and these are the hubs to which new nodes will prefer to attach as these hubs will be more efficient, more well connected, or stronger in a key facet. Unlike small world networks, scale-free networks display a power law distribution characterized by a small number of nodes with high degree and a large number of nodes with small degree measurements. The preferential attachment of new nodes in a power law distributed network reflects the concept of the "rich get richer." The hubs contribute substantially to the network's overall connectivity as they are the clumps of "glue" that bond the network together; however, little contribution is made by the more poorly connected nodes. Scale-free networks are more random in structure than lattice and small world networks, but less random than Erdös–Rényi (random) networks. The diameter of scale-free networks is smaller than other network architectures and this is because many nodes connect through the hubs.

3.5.6   Random (Erdös–Rényi) Networks

Random networks are characterized by new nodes joining the network at random, without any preferential attachment as would occur in scale-free networks. This means that the linking of new nodes occurs randomly with equal probability that a new node will link to any other node, hence random networks indicate a Poisson distribution. The degree centralities of nodes in a random network are not equal as in more regular topologies such as the lattice. Owing to the irregular nature of random graphs, they rarely reflect networks in the real world, as networks grow by nodes attaching themselves to the network; and it is commonly the case that nodes have a reason for joining the network and will chose to attach on the basis of some driving force. Substantial work on random networks has been carried out by Erdös and Rényi (1960) and Watts and Strogatz (1998). Figure 3.11 illustrates the structure of random networks. The disordered nature of the network can be easily seen together with a lower clustering of nodes than that found in other topologies. Random networks have the lowest clustering coefficient of all topologies and the effect of this is the low number of hubs, or highly connected nodes. It is relatively easy to navigate to opposite sides of the network because of the formation of random links. Each node in the random network contributes approximately equally to the network's connectivity.

3.5.7   Comparison of Network Topologies

The characteristics of networks summarized in Table 3.1 compare the major types of network structures discussed in the preceding text. The columns progressively become more random from left to right (lattice, which is highly structured, to random, which has a low structure). The ratings in each cell were compiled from numerous publications on network research, too many to list individually, and the findings for each criteria were not always clear or consistent across sources.

The stability of networks is a key factor for consideration in today's uncertain global world. Of particular interest is a network's ability to diffuse information to

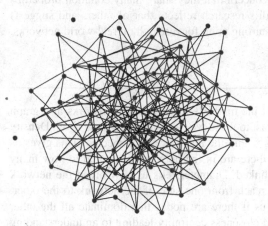

FIGURE 3.11   Scale-free network of 100 nodes.

its outer perimeters and also to withstand attack. Much of the research on diffusion relates to the adoption of innovations, which involves not only the spread of new ideas and devices but also the threshold at which the innovation is adopted. Diffusion in the context of this discussion on network topologies relates only to the spread of information and its cascading through the network to reach the peripheral nodes. While numerous studies have found that small world and scale-free structures diffuse information efficiently and quickly, research conducted by Noble, Davy, and Franks (2004) on the network structure and diffusion of innovation reported that the higher the proportion of random connections, the faster an innovative behavior spreads throughout the network. The heavily skewed distributions that result from preferential attachment appeared to slow diffusion down.

The research by Barabási et al. (2000) reported that scale-free networks are substantially more robust to the random removal of nodes than random graphs, but substantially less robust to the removal of hubs. Lewis (2009) studied the level of entropy in the different network topologies and concluded that scale free has less entropy (irregularity, chaos, or randomness) than random graphs, but more than small world or regular graphs such as the lattice. He also noted that conversely the small world graph is more structured than the random because its entropy is lower than that evident in all other classes except the regular lattice class.

However, the information in the summary table must be viewed with some caution. The results of the investigation of network topology in theoretical and simulated environments appear to differ depending on the size of the network, its density, and the way it is ordered. A number of researchers have studied the progression of graph topology from one side of the random scale to the other, gradually increasing or decreasing random node and link additions. Others take different approaches. Many researchers have found that real-world networks do not fit easily into the theoretical network topologies. Frequent findings are that real-world networks show pockets of a particular topology alongside sections showing another topology, commonly real-world networks appear to be a combination of several different network architectures. Other results (such as that of Airoldi, 2005) illustrate the difficulty in separating topological properties of cellular, core-periphery, and scale-free topologies and the concern that they share many common properties with the random topology. Airoldi's research reflects that of others and suggests that a mixture of types is a better starting point for research on real-world networks.

3.6 SUMMARY

In this chapter, we have looked at the methods for analyzing networks at a graph level. These measures are in contrast to those we calculated at a node level. Density will tell us how well all the nodes within the network are connected. This gives us an indication of how many links there are in the network compared to how many there could be if all nodes were linked. Diameter tells us how broad the network is, or how many steps it takes to reach from one side of the network to the other. Centralization measures will tell us if there are nodes that dominate all the other nodes in degree, betweenness, and closeness centrality leading to an understanding about the network structure. Average centralities will give us information regarding the average centrality measures for all the nodes in the network. We will then be able to ascertain the average degree, betweenness, closeness, and eigenvector centrality for nodes as these provide a piece of the overall puzzle that reflects the network. We also studied the structure of different network topologies and used graph level measures to compare networks. Together with the minimum and maximum centralities for the nodes the average centralities will tell us the range of node centrality measures, thus giving us more information about the network. Together, these graph level measures and topologies will permit us to gather valuable information about the network and permit us to see similarities and differences when comparing two or more networks.

Here is a summary of the key points discussed in this chapter:

- Graph level measures tell us about the structure and nature of the entire network, whereas node level measures tell us about the centrality of each node in relation to other nodes.

- The density of a network indicates how densely the nodes are connected through the entire network when compared with total possible connections. Sparsely connected networks will not be efficient in diffusing information across the network, whereas densely connected networks will spread information much more quickly owing to the increased number of links between nodes.

- Diameter of a network measures how many steps it takes to walk across the network from one side to another. This measure will assist us in determining how long it will take for nodes on the outer edges of the network to receive information and flows when compared to nodes closer to the center of the network.

- Degree centralization will indicate if there is one node that is very high in degree centrality while the other nodes are much lower in degree centrality. This standardized measure will be 0 when all nodes have the same degree centrality and 1 when one node is connected to every other node in the network.

- Betweenness centralization indicates if there is one node that dominates in betweenness centrality while the other nodes are low in betweenness centrality. A standardized measure of 0 indicates that all nodes have the same betweenness centrality and a measure of 1 indicates that one node is between all other nodes in the network.

- Closeness centralization will tell us if one node is very high in closeness centrality while all the other nodes are low in closeness. A standardized measure of 0 will tell us that no one node is close to all other nodes and a measure of 1 indicates that one node is close to all other nodes in the network.

- Average centrality measures for nodes gives us limited information on its own; however, coupled with minimum and maximum node centrality measures, or combined with centralization measures, average centralities can provide valuable information when analyzing the structure of the network.

- Calculation of these measures via an automated tool can be achieved and Lab Exercise 3 walks the reader through this process using the ORA software provided. This can also be calculated using other automated tools.

CHAPTER 3 LAB EXERCISE

Network Level and Centralization Measures using the Organizational Risk Analyzer (ORA) Software System

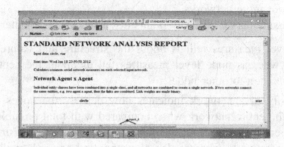

Learning Objectives

1. Determine and explain average centralities for a network in ORA.
2. Determine and explain network density and diameter in ORA.
3. Determine and explain network centralization measures in ORA.
4. Learn how to generate and visualize .different types of networks using ORA

1. Determine Average Centralities for Networks in ORA

Run ORA and select the *Add New Meta-network* icon. Name the network "star." Select the *Add New Class Node* icon and add six Agent nodes. Select the Add New Network icon and add an *Agent × Agent* network.

Using the image in Figure 3.4a, enter the corresponding adjacency matrix entries for this network. This is an undirected network, so all the links are reciprocated.

Save the meta-network by selecting *File → Save Meta-Network As →*, then save the meta-network with the name *star.xml*.

Using the same steps, repeat the exercise and enter the circle network in Figure 3.4(b). Save the meta-network with the name *circle.xml*.

Run a Standard Network Analysis report for each network by selecting the *Generate Reports* tab and scrolling down to *Standard Network Analysis*. Select All meta-networks so that the star and circle meta-networks are ticked. Use the default parameters and produce in HTML format. (In the previous lab exercise, you generated an All Measures report. Network level measures are also shown in the All Measures report. Select Generate Report, select All Measures. Use default parameters and produce the report in HTML. Select the Analysis for network *Agent* × *Agent* link.)

Let us go back to generating the Standard Network Analysis report:

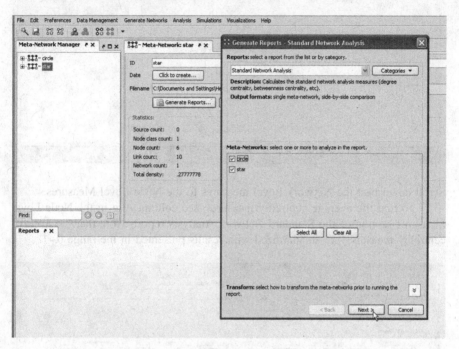

Your report for the star and circle networks should look like this:

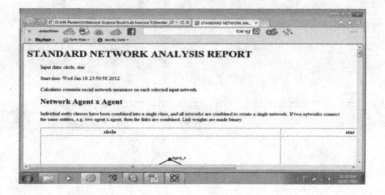

Scroll down to see the network graphs:

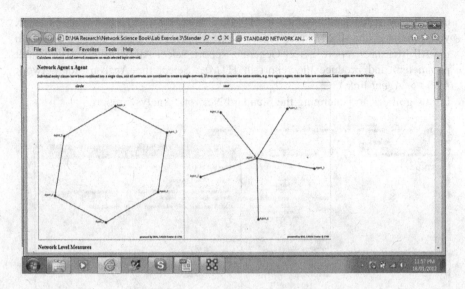

Scroll down past the Network Level measures to the Node Level Measures.

To find the average centrality measures, see column Avg in the Node Level Measures section of the Standard Network Analysis report. Note that the average centrality measures are standardized with results presented in the range 0–1.

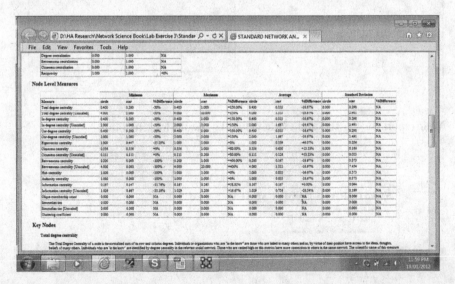

Let us compile a summary comparison table containing the average total degree, average closeness, and average betweenness:

Measure	Circle	Star
Average total degree	0.400	0.333
Average closeness	0.556	0.630
Average betweenness	0.200	0.167
Average eigenvector	1.000	0.539

Can you explain the differences in the average centralities? Here is a brief explanation:

Average total degree for the circle graph is 0.400 and is based on all nodes having the same degree centrality (each node is connected to exactly two other nodes), whereas the star graph has an average total degree of 0.333 owing to the central node that is connected to all other nodes. Both have less than half the possible links in the network.

The average closeness for the star network is 0.630 and is greater than the circle network at 0.556 as some nodes are further apart in the circle network, whereas all nodes are either one or two steps from any other node. The average betweenness for the circle network is 0.200, which is slightly greater than that for the star network at 0.167 as more nodes lie on the shortest path between two other nodes in the circle network.

The average eigenvector for the circle graph is the maximum 1.000 as all nodes are linked to other influential nodes. In the star network, the nodes are linked via Agent_4 in the center of the graph reflected by an average eigenvector value of 0.539.

Now enter the data for the Dolphins network from Table B.6 and Table B.7 in Appendix B. Save the meta-network as dolphins.xml. You can also download this network data from the CASOS web site. This network illustrates the relationships between 62 dolphins studied in Doubtful Sound in New Zealand. The relationships plotted are individual dolphins that were seen together more often than expected by chance. For more information you can read two publications by David Lusseau that explain the research on this network of dolphins (see the references at the end of this lab exercise).

Produce a Standard Network Analysis Report for this network. Your report should look like this:

STANDARD NETWORK ANALYSIS REPORT

Input data: dolphins

Start time: Fri Jan 06 20:08:48 2012

Calculates common social network measures on each selected input network.

Network agent x agent

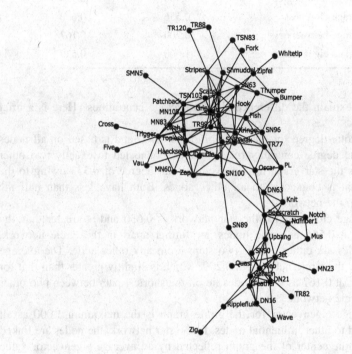

Network Level Measures

Measure	Value
Row count	62.000
Column count	62.000
Link count	318.000
Density	0.084
Isolate count	0.000
Component count	1.000
Reciprocity	1.000
Characteristic path length	3.357
Clustering coefficient	0.259
Network levels (diameter)	8.000
Network fragmentation	0.000
Krackhardt connectedness	1.000
Krackhardt efficiency	0.946
Krackhardt hierarchy	0.000
Krackhardt upperboundedness	1.000
Degree centralization	0.116
Betweenness centralization	0.212
Closeness centralization	0.227
Reciprocal?	Yes

Now let us add the results for the dolphin network to our table:

Measure	Circle	Star	Dolphins
Average total degree	0.400	0.333	0.084
Average closeness	0.556	0.630	0.307
Average betweenness	0.200	0.167	0.039
Average eigenvector	1.000	0.539	0.287

We can see that the dolphin network has a lower total degree than the circle and star networks as it has a smaller proportion of possible links in the 62-node network. The size and nature of the network are also reflected in the other average centrality results, which are lower than those for the other two networks. Many dolphins have no contact whatsoever with other dolphins in the network and several key dolphins provide the connections between smaller groups within the pod. These connecting dolphins are known as boundary spanners as they are the nodes that link groups together. We will look at grouping and clustering in a future exercise.

2. Determine Network Density and Diameter

Find the network density and diameter for the three meta-networks we have loaded and add these to our table of results.

Measure	Circle	Star	Dolphins
Average total degree	0.400	0.333	0.084
Average closeness	0.556	0.630	0.307
Average betweenness	0.200	0.167	0.039
Average eigenvector	1.000	0.539	0.287
Density	0.400	0.333	0.084
Diameter	3.000	2.000	8.000

The density of the network is the same as average degree centrality at the graph level. Density measures how well nodes are connected to one another based on the definition of a link. A density of 0.084 means that the network is sparse, with the number of links between nodes being only a small percentage of the possible total number of links if all nodes were linked to all other nodes. This means that flow through the links will be limited to only those connections that exist, and in this case the number of links is small.

The diameter of the network describes how much effort is involved in moving from one end of the network to the other and is represented by the longest geodesic in the network. For the star graph, the maximum geodesic is length 2, thus the diameter = 2. In the star network, Agent_4 is at the center of the graph and connects to all the other nodes. Agent_4 can easily and quickly pass information to the other nodes in just one step and no node in the network is more than two steps away from another node. Agent_4 is a broker in this network. For the circle graph, the diameter = 3. In this graph there is no brokering node, so it takes longer for information to travel from one node to other nodes in the network. The diameter of the dolphin network is 8. This is because it is a larger network than the circle and star and it is not as compact. Hence, it takes longer for information to travel to the outer parts of the network.

3. Determine Network Centralization measures in ORA

Using the All Measures Report and/or the Standard Network Analysis report find the centralization measures for each of the three networks and add these to our table:

Measure	Circle	Star	Dolphins
Average total degree	0.400	0.333	0.084
Average closeness	0.556	0.630	0.307
Average betweenness	0.200	0.167	0.039
Average eigenvector	1.000	0.539	0.287
Density	0.400	0.333	0.084
Diameter	3.000	2.000	8.000
Degree centralization	0.000	1.000	0.116
Betweenness centralization	0.000	1.000	0.212
Closeness centralization	0.000	1.000	0.227

Centralization measures indicate nodes that are special in that they are more central to the network than other nodes in some way. The three centralization measures—degree centralization, betweenness centralization, and closeness centralization—indicate whether an individual node has a much higher centrality than a typical node. These measures are standardized between 0 and 1. A measure of 0 indicates that no node exceeds the others (i.e., all nodes are the same) and a measure of 1 indicates one node dominates all others.

Degree centralization will indicate nodes that dominate all other nodes in the network. Removing key nodes in a network with large dominant nodes will result in division of the network into smaller groups of nodes as removal of these key nodes also removes the links, thus breaking the network apart. A measure of 1 for degree centralization indicates that one node dominates all other nodes. A measure of 0 indicates that all nodes are the same in degree centrality. Networks with low degree centralization do not have a high risk of fragmentation as there are no exceptional nodes that connect to all other nodes in the network.

Betweenness centralization indicates the presence or absence of nodes that are high in betweenness and act as gatekeepers in the network. A between centralization of 1 indicates that one node controls access to all other nodes and is most often on the shortest path between two other nodes in the network. A betweenness centralization of 0 indicates that all nodes have the same betweenness centrality.

Closeness centralization indicates whether one node is high in closeness centrality to the exclusion of others. A closeness centralization measure of 1 indicates that a central node, which is one step away from every other node in the network, exists. This central node is key to passing information to the other nodes in the network, as all other nodes are a minimum of two steps away from each other. A closeness centralization of 0 indicates that every node passes information to every other node in the same number of steps.

Looking at the results in the Standard Network Analysis report, the circle network has 0.000 for degree, betweenness, and closeness centralization. This means that all nodes have the same degree, betweenness, and closeness centralities, and no one node is more central in the network. The star network, on the other hand, has 1.000 for degree, betweenness, and closeness centralization. This means that there is one node that is totally central to the network it dominates, in degree centrality connecting to all other nodes, betweeness centrality as it lies between all other nodes as the gatekeeper in the network, and closeness centrality being only one step away from all nodes and every node being at least two steps away from every other node.

The centralization measures for the dolphin network are between those for the star and ring networks, which are the two extremes in centralization. The degree centralization is 0.116, which indicates that there are very few nodes that are high in degree centrality when compared to the other nodes in the network. Betweenness centralization is 0.212, indicating that there is no one single node that acts as a boundary spanner or gatekeeper controlling access to all other nodes in the network. Closeness centralization is 0.227, again indicating that there is little evidence of one dominant node: there is no one node that is only one step away from every other node. The nodes in this network have different closeness centrality measures as can be seen by the details in the Avg, Min/Max and Min/Max Nodes columns in the Analysis for network agent × agent section of this report. You can also see the difference in these measures for each node in the Agent-Level Measures section of this report.

Generate Networks in ORA

The objective of this next task is to familiarize you with the generation and visualization of different types of network topologies using the ORA software. As you move through the exercise, note that the networks you generate in this lab will be slightly different than the examples in the book. The networks in this lab are generated via stochastic algorithms. It is important that you understand the characteristics of Erdös–Rènyi, core-periphery, scale-free, cellular, lattice and small-world networks.

Load ORA

You are now going to generate some example networks. Click on the Generate Networks tab on the top left of the ORA screen, then choose Create Stylized Network. Let us start with an Erdös–Rènyi, random network.

Generate Erdös–Rènyi Network

Enter the following information in the popup window. Create a new meta-network with ID: Erdös–Rènyi.

Choose the default settings as follows:

Source node class—Create the new class: is checked, type: Agent, id: Agent, size: 100.

Target node class—Create the new class is checked, type: Agent, id: Agent, size: 100.

Density should be 0.05, Allow diagonal links is unchecked, and Create a symmetric network is checked. Enter an output network ID: Erdös—Rènyi network

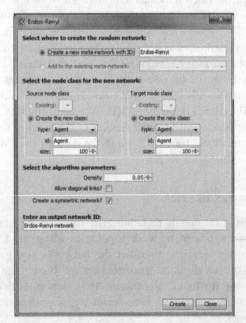

Press Create at the bottom of window. Then you must press Close to close this window.

Erdös will now be listed in the Meta-Network Manager section and appear as the Meta-network Erdös–Rènyi. Choose the + on the left of the meta-network name to expand the file to list the node types and networks.

Let us take a look at this network. Select the tab *Visualize*.

Your network will look something like this, but not exactly the same, as each randomly generated network will be slightly different! Toggle the show labels (key ring) tab to show or hide the node labels.

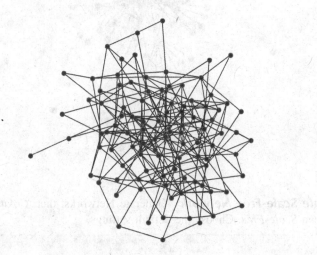

Now save your image, select *File*, then Save Image to File.

Enter File Name: *Erdöos Network Example* and save in JPEG format.

Generate Core-Periphery Network   Use the same procedure to generate the following networks using the default settings. The images will not be exactly the same, as these networks are randomly generated. Use the network type as the network ID.

Generate Networks, then *Create Stylized Network*, then *Core Periphery*. Choose the default settings.

Your Core-Periphery network should look like this:

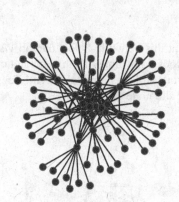

Generate Scale-Free Network   Generate Networks, then *Create Stylized Network*, then *Scale-Free*. Choose the default settings.

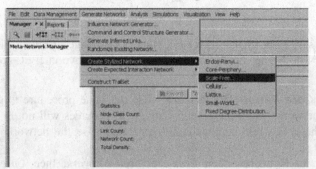

Your Scale-Free network should look something like this:

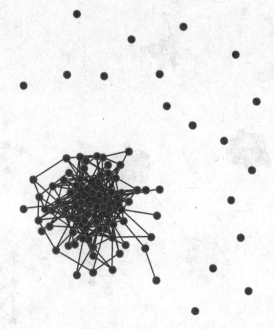

Generate Cellular Network   Generate Networks, then _Create Stylized Network_, then _Cellular_. Choose the default settings.

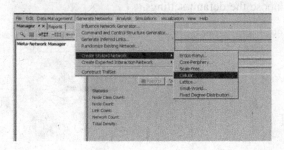

Your Cellular network should look something like this:

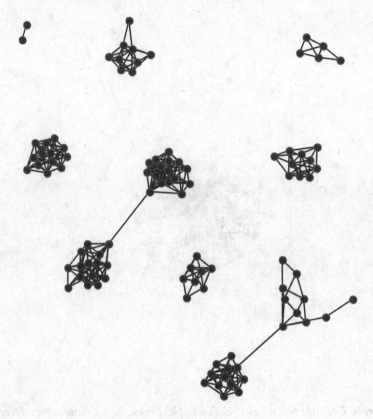

Generate Lattice Network Generate Networks, then _Create Stylized Network_, then _Lattice_. Choose the default settings.

Your Lattice network should look something like this:

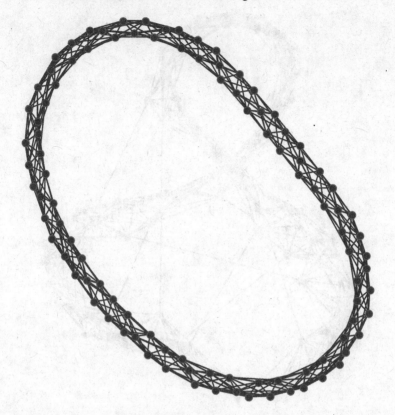

Generate Small-World Network   Generate Networks, then *Create Stylized Network*, then *Small World*. Choose the default settings.

Your Small-World network should look something like this:

Build Comparative Table Develop a table comparing the values for graph density, cluster coefficient, network diameter, average path length, and diffusion and network centralization measures. Use the following format to guide your analysis. These measures can be found by generating the All Measures report for each network in ORA. Explain your findings.

Measure	Erdös-Renyi	Core-Periphery	Scale-Free	Cellular	Lattice	Small-World
Density						
Cluster coeff.						
Diameter						
Avg. path length						

(continued)

Measure	Erdös-Renyi	Core-Periphery	Scale-Free	Cellular	Lattice	Small-World
Diffusion						
Network centra. degree						
Network centra. degree						
Network centra. between.						
Network centra. closeness						
Network centra. eigenvector						

Summary

In this lab exercise, we have produced reports giving us details of average centralities, diameter and density, and network centralizations for three networks. These measures give us some tools for understanding the structure of networks and determining whether there exist any key nodes in the network. They also give us a means of comparing networks as a whole.

Load up more networks of your choice and perform the same activities. Look at the network graphs and the reports to become more familiar with these measures in different network situations.

EXERCISES

Generate the network graphs and adjacency matrices for five networks, all with 25 nodes, but differing topologies. All five networks are undirected. Complete the table by calculating the measures shown either manually or using an automated tool and answer the questions.

3.1. Determine the density and diameter of each network. Explain the difference between the networks.

3.2. Calculate the average degree, betweenness, closeness, and eigenvector centrality for each network. Compare these and explain the difference between the networks.

3.3. Calculate the network centralization for degree, betweenness, and closeness for each network. Compare these and explain the difference between the networks

3.4. How difficult would it be to fragment each network and how could this be done?

Measure	Scale free	Small world	Core periphery	circle	star
Average total degree					
Average closeness					
Average betweenness					
Average eigenvector					
Density					
Diameter					
Degree centra.					
Betweenness centra.					
Closeness centra.					

REFERENCES

Airoldi, E. and Carley, K. (2005). Sampling algorithms for pure network topologies: a study on the stability and the separability of metric embeddings. *SIGKDD Explorations Newsletter*, **7**(2):13–22.

Barabási, A.-L. and Albert, R. (1999). Emergence of scaling in random networks. *Science*, **286**(5439):509–512.

Barabasi, A.-L., Albert, R., and Jeong, H. (2000). Scale-free characteristics of random networks: the topology of the world-wide web. *Physica A: Statistical Mechanics and its Applications*, **281**(1–4):69–77.

Borgatti, S. and Everett, M. (2000). Models of core/periphery structures. *Social Networks*, **21**(4):375–395.

Carley, K., Lee, J., and Krackhardt, D. (2001). Destabilizing terrorist networks. *Connections*, **3**: 79–92.

Dodds, P., Muhamad, R., and Watts, D. (2003). An experimental study of search in global social networks. *Science*, **301**(5634):827–829.

Frantz, T. and Carley, K. (2005). A formal characterization of cellular networks. CASOS Technical Report CMU-ISRI-05-109. Center for Computational Analysis of Social and Organizational Systems.

Frantz, T., Cataldo, M., and Carley, K. (2009). Robustness of centrality measures under uncertainty: Examining the role of network topology. *Computational and Mathematical Organization Theory*, **15**(4):303–328.

Freeman, L. (1979). Centrality in social networks conceptual clarification. *Social Networks*, **1**(3):215–239.

Granovetter, M. (1973). The strength of weak ties. *American Journal of Sociology*, **78**(6):1360–1380.

Kleinberg, J. (2000). Navigation in a small world. *Nature*, **406**(6798):845–845.

Kleinberg, J. (2001). Small-world phenomena and the dynamics of information. *Advances in Neural Information Processing Systems (NIPS) 14*.

Krebs, V. (2002). Mapping networks of terrorist cells. *Connections*, **24**(3):43–52.

Latora, V. and Marchiori, M. (2001). Efficient behavior of small-world networks. *Physical Review Letters*, **87**(19):198701.

Lewis, T. (2009). *Network Science: Theory and Applications*, 1st edition. Wiley.

Travers, J. and Milgram, S. (1969). An experimental study of the small world problem. *Sociometry*, **32**(4):425–443.

Watts, D. (2003). *Small Worlds: The Dynamics of Networks between Order and Randomness (Princeton Studies in Complexity)*. Princeton University Press, illustrated edition.

Watts, D. and Strogatz, S. (1998). Collective dynamics of 'small-world' networks. *Nature*, **393**(6684):440–442.

PART II

SOCIAL THEORY

CHAPTER **4**

SOCIAL LINKS

> At the same time, as social beings, we are moved in the relations with our fellow
> beings by such feelings as sympathy, pride, hate, need for power, pity, and so on.
> —Albert Einstein

Learning Objectives

1. Understand and describe the concept of social circles and how they are formed.

2. Understand and describe the concept of link formation and the influence of Homophily, Reciprocity, Proximity, Prestige, Transitivity, and Balance in link formation.

3. Understand the following concepts and describe their applicability in social networks:

   (a) Social Conformity.

   (b) Structural Holes.

   (c) Social Capital.

   (d) Link Optimization.

   (e) Clusterability.

Why Do We Connect?

There exist a certain set of social dynamics that dictate how people form relationships. Social network analysis provides a context in which to understand how these forces work upon individuals. The framework of a hierarchy of social link motivation presented here is straightforward, designed by the authors, and serves to organize the vast research on social theory that drives link formation in networks.

Social Network Analysis: with Applications, First Edition.
Ian A. McCulloh, Helen L. Armstrong, and Anthony N. Johnson.
© 2013 John Wiley & Sons, Inc. Published 2013 by John Wiley & Sons, Inc.

109

4.1 INDIVIDUAL ACTORS

Fundamentally, people need to feel valuable and this feeling is only validated in the presence of others. Cooley (1902) refers to this concept as the "looking glass self." There are three components to this theory: (i) people consider how others perceive them, (ii) they imagine how others judge that appearance, and (iii) they develop self-identity through the perceived judgment of others (Yeung and Martin, 2003). The very activities that people pursue, the accomplishments that they take pride in are those that other people recognize as valuable. Psychologist Erik Erikson's theories on personality and identity have self-worth at the core of all stages of human development (Erikson, 1956; Marcia, 1966). Carl Rogers (1961) theories on personality and the self-concept are examples of social group validation of individual identity and self-worth. In essence people need to belong to a social group.

People are constrained, however, in the number of relationships they can maintain. British anthropologist Robin Dunbar (2010) suggested that there existed a cognitive limitation on the number of stable relationships a person can maintain. This did not include people who the individual knew in the past but had ceased a regular relationship with, nor did it include latent friends, where there was no regular contact. She used a model connecting the neocortical ratio to primary group size to predict that humans should have about 150 individuals in their primary social group. The neocortical ratio is the volume of the neocortex divided by volume of the remaining brain. Dunbar did not provide a precise definition of relationship. Killworth and Bernard (1979) conducted several empirical studies, which have shown this limit to be as high as 290 connections with a median of 231 links, which is based on empirical findings of observed human relations. Mayhew and Levinger (1976a, 1976b) show findings that suggest that the density of connections between actors is a function of the context of the relationship and the size of the population. Intuitively, we can also recognize that an individual has a finite amount of time in the day to invest in relationships. Thus, the number of links an actor maintains is a function of three main variables: the nature of the relationship (online, acquaintance, face-to-face friendship); the required amount of time invested to define that a relationship exists; and the cognitive limitations of the actor.

The need for self-worth to be validated by others and the actor's constraints in terms of time and cognitive limitations drive individuals to maximize the utility of their relationships (Lospinoso et al., 2009). They may choose to establish many low value ties, few high value ties, diverse ties across different social groups, among other strategies. This has several important implications for link formation in networks. First, individuals may derive different utility from different networks. Some networks may provide access to knowledge or resources, which make them valuable in another setting. Other networks may provide social entertainment. Different networks may provide love and security. Yet other networks may provide communities of interest for different hobbies, activities, and employment. Charles Kadushin (2012) identified these different networks as *social circles*.

Most individuals have membership in multiple social circles, and it is rare that two individuals will have all of their social circles in common. Feld (1981) explains this as Focus Theory. He describes social context as a set of foci and individuals. Individuals are related to the extent that they share a focus. Feld considers a focus to be a "social object," which could be group membership, a work place, ethnic identity, attitude, and so on. He proposes that individuals' shared relations to foci create social circles and increase their likelihood for interaction. This is an explanation for overlapping patterns of network clusters. Maintaining membership in multiple social circles is also efficient from an optimization perspective. By maintaining relationships with a few individuals from disparate groups, an individual can have access to a wider variety of resources without the cost of establishing relationships with all members of the group.

4.2   SOCIAL EXCHANGE THEORY

Another key dynamic for network link formation is social exchange theory. People in positions of advantage within the network often exploit their position for advantage. Exchange theory was first presented by Homans (1958) and it suggests that all human relationships or social interactions are exchanges based upon an analysis of costs and benefits as well as analysis of alternatives. Social exchange theory evaluates the social associations analyzed in terms of the value of giving and receiving, and when the costs outweigh the benefits then the relationship breaks down.

The exchange process is complex as relationships can be interdependent and are context driven. Zafirovski (2005) observes that reciprocity compels exchange in social interaction, and relations do not continue where reciprocity is breached, as relations continue based upon the expectation that the interaction will be mutually beneficial. Blau (1994) believes the initial drive for social interaction is initiated by the perceived exchange of benefits, both intrinsic and extrinsic, regardless of cultural norms. Molm (1990) discovered that theories that neglect the effect of actors within given structural conditions are incomplete, and how much power an actor has and how it is used to respond to another's behavior are distinct determinants of exchange outcomes. Later research (Molm et al., 2000) on negotiated and reciprocal direct exchange showed that reciprocal exchanges produced stronger trust and commitment than negotiated exchange, and that behaviors indicating or confirming the partner's trustworthiness had a greater influence upon trust in the reciprocal exchange.

Emerson (1969) posits that reciprocity in social exchange links closely with the concepts of power, dependence, and cohesion. Power is derived from a position to control resources and hence wealth and is the property of a relation rather than an actor, because it is inextricably linked with the other's dependency (Emerson, 1962). It is not difficult to see how power can influence the outcomes of exchanges and stability of the associated networks. As networks can be combinations of complex relationships much research has been conducted at basic levels of exchange on very small networks in an effort to understand what happens at a more cellular

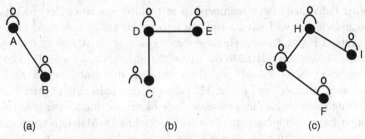

FIGURE 4.1   Power in basic network exchanges: (a) two-node network, (b) three-node network, and (c) four-node network.

level. A number of nodes are required to negotiate the split of given resources, commonly money. Figure 4.1 illustrates simple networks of two to four nodes in an exchange activity.

The results are reasonably simple in a network of two nodes, say nodes A and B. The nodes negotiate and divide the resource, and the result is most commonly an equal split. In a three-node network consisting of nodes C, D, and E in Figure 4.1, node D has greater power as it is positioned between C and E and is able to negotiate with both these nodes. However, nodes C and E are only able to negotiate with node D. In the four-node network consisting of nodes F, G, H, and I, nodes G and H hold the greatest power as G can negotiate with F and H, and node H can negotiate with G and I. If any of the four nodes have a resource to share then nodes G and H are in the most favorable positions and can exert the most power on individual exchanges. They also stand to obtain the best outcomes. When you add other influences to relationships between nodes in a network, such as an organizational reporting responsibility, a homophilous belief or a reciprocal expectation, it becomes more difficult to isolate and quantify the effect of each influencing factor.

Check on Learning

1. Briefly explain the concept of "social circles."
2. Briefly explain the concept of "social exchange theory."

Answer

1. Individuals seek confirmation of self-worth and different social networks provide the needed validation. An individual's needs will provide the impetus for the nature and strength of ties and their diversity of associations.
2. Social exchange theory evaluates links between people based upon the value of giving and receiving, and the exchange will continue as long as the benefits are greater than the costs.

4.3   SOCIAL FORCES

There are six key social forces that drive the formation of social links: homophily, reciprocity, proximity, prestige, transitivity, and balance. All of these drivers of link formation represent costs and benefits associated with establishing and maintaining links in a network.

4.3.1   Homophily

Homophily deals with the tendency of individuals to form relations with those like themselves (Blau, 1977; McPherson et al., 2001). If two actors share common interests, beliefs, goals, race, gender, and/or culture, they are more likely to form connections than if they shared no common interests or features. This is often commonly termed *birds of a feather flock together*. Homophily increases the likelihood that alters will have values that reinforce the individual's self-worth. As people have multiple attribute characteristics, homophily must be evaluated across multiple dimensions, known as *Blau-space* (McPherson, 1983), named after the scientist who proposed the concept in 1977. Homophily is one of the most robust findings in social science.

Interesting studies have explored whether perceived homophily exists in networks, that is, peoples' perceptions of others' interests and beliefs drive their behavior. With perceived homophily, an individual's perceptions are updated through interaction, which may lead to dynamic changes in network behavior over time. Interaction can also create opportunities to modify ideology and value systems, which can contribute to social conformity and group think as actors become more similar in an unconscious effort to seek greater validation of self-worth.

4.3.2   Reciprocity

Reciprocity is whether agents tend to form directed relationships with alters who initiate relationships with them. In other words, if agent A sends an email to agent B, does agent B tend to send emails back to agent A? From a utility perspective, it is easier to maintain a social relationship when the other actors take an active role in reaching out and reciprocating the relationship. Reciprocity may partially be an emergent property of homophily if two similar agents develop a reciprocal relationship due to their shared attributes.

Reciprocity has become a key aspect of social networking using social media and communications such as Facebook and Twitter blogs. Status is seen to be gained from the number of reciprocal contacts and conversations carried out and the number and prestige of connections are important for the perceived prestige of the individual. Communication between two peers would be expected to follow the following interaction format: user A sends user B a message and receives a reply, and the continuing exchange will be bidirectional and symmetric. However, if user A were to send multiple messages to user B who is a celebrity, it is highly unlikely that B will reply and the interaction is unreciprocated. Cheng and colleagues (2011) formulated an approach to predicting the symmetry of the interaction using node attributes that approximate the relative status of the two communication nodes. In closely knit groups based upon homophily the lack of reciprocity can weaken the ties between group members, making the group more fragile as members do not abide by the expected norms of the group.

4.3.3   Proximity

Proximity is the organizational or physical distance between nodes. This can be organizational or physical. Organizational proximity refers to a semantic distance between groups in an organization. People that work together in shipping and receiving are closer in organizational proximity than they are to a person in manufacturing. They may have an even greater distance to a person in accounting, even if that section is physically closer to the shipping and receiving office. Within an organization, there are often common meetings, goals, and joint tasks, which provide opportunities for interaction. An individual in a different organization will often have conflicting demands. This phenomenon makes it easier for members of the same organization to interact than for members of different organizations. Thus, organizational distance creates a cost to link maintenance.

Physical proximity is the actual physical distance between individuals. Sailer and McCulloh (2011) showed that this was also based on perceptions. Individuals in an open office plan perceive distance based on how visible others are within the office, whereas with closed individual offices a metric distance equivalent to the number of paces an individual walks to interact with another is more appropriate. In either case, the closer actors are to one another, the more likely they are to interact and form relationships. Greater physical distance incurs a cost of link maintenance in terms of time to reach another node or in terms of the technological effort, such as a phone or email.

As groups form, social norms are established. A norm is a value of appropriate or inappropriate placed upon feeling, thought, or behavior. Friedkin (2001) suggests that shared belief in these values with influential others reinforces and validates the belief, thus creating the group norm. He further suggests that these norms provide an informal set of expectations, which provide opportunities for individuals to validate their sense of self-worth within the context of the group or social circle. An individual's choice to conform with the group is driven by their need for social acceptance and validation and a perception of how well the group's norms align with their positive attributes. Group membership can form an integral part of a person's individual identity.

4.3.4   Prestige

The establishment of social norms provides a mechanism for two important concepts in social psychology, *prestige* and *social conformity*. Individuals who epitomize the social norms and values of the group not only derive a personal sense of self-worth, but also are perceived by others to be valuable. These individuals have high prestige and hold greater influence over the attitudes and ideology of group members. They are also able to influence group norms that may emerge in the future. Other members of the group will often choose to link to high prestige nodes for greater validation of their own self-worth.

Perceived motivation can influence individual status within the group as well (Ridgeway, 1978; 1982). Low status individuals who are perceived to be group-oriented combined with reasonable task proficiency tend to attain higher status levels over time. By contrast, individuals who are low status and are perceived to be self-oriented tend to remain low status. High status individuals tend to remain high status regardless of their motivation as long as their task proficiency remains high.

Prestige can also be achieved in terms of an individual's access to resources, knowledge, and other social circles. Within this context prestige can be deliberately improved. A node can strategically develop ties between social circles that will benefit one another, thereby increasing their value. They can position themselves within close proximity to others in the network, increasing their opportunities for social interaction. Individuals can demonstrate an adoption of group values making themselves more homophilous to others in the group. Finally, simply being outgoing may leverage the concept of reciprocity by reducing the link maintenance cost of others.

The concept of prestige enables high prestige nodes to maintain more links with less cost. As group members have a desire to connect to a high prestige node, they are willing to overcome other social costs such as a lack of proximity, reciprocity, or homophily, although nodes that exhibit low social cost are even more attractive. This does not mean that a prestigious node is unconstrained, however. They still must make some effort to maintain the relationship from their end. These nodes also must maintain social norms in order to retain their prestige within the group. This can sometimes be challenging if prestigious individuals choose to affiliate with social circles that may have competing social norms.

4.3.5 Social Conformity

Individuals who do not adopt group norms will have reduced opportunity to validate their sense of self-worth. Conforming to the group norms provides them an opportunity to validate their sense of self-worth. This mechanism drives *social conformity*.

There is a social network group effect that contributes to conformity. That is, research shows repeatedly that groups can exert implicit pressure to influence opinions leading to higher conformity, more extreme views, and a group think mentality where contradictory evidence is ignored. For example, Solomon Asch (1955, 1956) conducted a study, where subjects were placed in a group, where the other group members were given information that was different from reality. For example, eight individuals (the subject and seven others) were shown three lines of various lengths and asked to verbally identify which line matched a given fourth line. The subject was positioned to answer after the seven other individuals. Most of the time, the seven other individuals would give a correct response, but in certain cases, the other subjects would agree on a wrong answer. Asch then measured how the subject would modify their answers based on the majority opinion. The objective of the experiment was to understand factors that would enable people to resist group/peer pressure.

The study found that 37% of the subjects conformed to the wrong responses given by the majority. It was not clear if the subjects actually believed their incorrect response (*informational conformity*), or if they were simply conforming to the group so they were not singled out (*normative conformity*). A variation of this experiment had one of the seven other individuals give correct responses, and the conformity dropped from 37% to 5.5%. These findings suggest that an individual's perception of reality, to include their attitudes, may be significantly influenced by other people to whom they are socially connected.

McCulloh (2013) conducted a social network variation of Asch's experiment using a platoon in the US Army. Social network data was collected on the platoon, recording directional ties of self-reported friendship and respect, creating two sets of network data on the same subjects. Four nodes that were highly central in both networks and four nodes that were peripheral in both networks were selected to be the subject of an Asch type experiment. Eight other nodes in the network were selected to serve as confederates of the experiment. Instead of asking participants to offer opinions on line length, the individuals were told that they were preparing for promotion boards and were being tested on military knowledge. Some of the questions were modified to offer an obviously wrong answer. The most ridiculous question was, "when do you apply a tourniquet to a neck wound? (i) when it is spurting blood; (ii) when it is coming from an artery; or (iii) never."

This study demonstrated a profound network effect on conformity. Central nodes conformed to the group and provided wrong answers 7% of the time on average. Peripheral nodes, however, conformed 62% of the time. The strongest correlation between centrality and the number of wrong answers occurred with betweenness centrality in the friendship network with a correlation of −0.84.

This study provides evidence of the need for social acceptance. Individuals who are peripheral and have need of social acceptance may conform as a way to gain group acceptance. Central actors who already have a sense of acceptance are free to identify the "wrong" answers without fear of group sanction.

Mcculloh et al. (2010) studied the social network effects on post-traumatic stress disorder (PTSD). They recorded data on the mental health of 1000 soldiers in a US Army brigade at three time periods; 2 months prior to deployment, 3 months into their deployment, and 2 months following their 1 year deployment to Afghanistan. They found that peripheral individuals in the friendship network were more likely to exhibit symptoms of PTSD and depression. In a similar series of studies on isolated groups at the South Pole, Jeff Johnson et al. (2003) discovered that peripheral individuals would often develop thyroid problems, which is related to depression. Thus, individuals may have a physiological need for social acceptance and conformity is a defense mechanism to achieve that acceptance.

In the famous Milgram Experiment (Milgram 1963, 1974) conducted at Yale University, Milgram studied the willingness of people to obey perceived authority figures, when their instructions conflicted with moral conscience. He found that 26 of the 40 subjects (65%) followed the immoral instructions. This suggests that people can be influenced to a greater degree, by a minority of people, if the minority is seen as authority figures. These authority figures derive their authority informally from epitomizing group norms, as well as formally. Thus nodes with high centrality may serve as an opinion leader for the group.

At the point where people have joined various social circles, derived some benefit from their associations, and potentially conformed to group norms, they are in a position to optimize the utility they derive from their social circles. There are several mechanisms for this behavior: transitivity, social capital, link optimization, and balance. These may not be conscious mechanisms, but they are powerful social forces that describe many dynamics of informal networks.

4.3.6   Transitivity

Transitivity of relations means that if there is a link from actor A to actor B, and a link from actor A to actor C then there is a tendency for actor B and actor C to form a link with each other. In other words, connections are formed through a common connection to a third party. The underlying social mechanism for transitivity may overlap competing social forces. If homophily has driven the links between actors A and B and between actors A and C, then there would be similar attributes that would drive the connection between actors B and C in a homophilous relation. On the other hand, if the connection involves time spent in social interaction, actors B and C would have greater opportunities to meet through their shared proximity to actor A. Therefore there are three potential confounded explanations behind transitivity. Transitivity may be an emergent network property based on (i) homophily, (ii) proximity, or (iii) brokered social relations from a common friend or acquaintance.

Modeling transitivity in statistical models and simulations can present major problems, which argue in favor of the emergent property perspective. If actors attempt to establish transitive links in a network, then the tendency would be for networks to have one giant fully connected component and many isolate nodes with no connection to the network. This usually does not represent observed network behavior. These models are degenerate in that the empirical data is extremely improbable under the models estimated from the same data.

An improvement to the concept of transitivity is geometrically weighted edgewise shared partners (GWESP) (Hunter and Handcock, 2006). This approach models a diminishing marginal return in the utility for additional shared transitive links. In other words, if two actors share a connection to a common alter, they are more likely to establish a link. If they have three alters in common they are even more likely. Four alters are more likely still. However, if they have 20–30 alters in common and have not established a link, there is probably a good reason for it. Thus, each additional shared alter is less important in terms of utility. The rate at which this decays is a geometric parameter in the GWESP model. This approach has much better modeling properties. It is rarely degenerate and thus supports the theory of utility as a mechanism of establishing social links.

Sometimes transitivity can be deliberately employed by an actor to increase their utility within the context of one or more social circles. One of the benefits of social relations is the access to knowledge and resources. Nodes that occupy a network position in between a key resource and others in the network derive informal power as discussed in Chapter 2. However, they also incur a social burden and cost to maintain relationships with the two groups (those with the resource and those requiring the resource). Brokering an introduction can provide an actor the opportunity to derive a sense of self-worth and value by facilitating a group's access to the resource, without the requirement to maintain the links, especially if the volume of resource flow is large (Feld, 1981).

4.3.7   Balance

Links do not always represent positive affinity and in some cases negative affinity may affect positive social relations. *Balance* (Heider, 1946; Cartwright and Harary, 1956) is an aggregate level measure that tests whether "the enemy of my enemy is my friend." If two actors that are "friends" have the same affinity to another there is balance. If the two friends have differing affinity to another there is a cognitive dissonance. *Cognitive dissonance* (Festinger, 1957) is a discomfort resulting from holding conflicting views at the same time. For example, a person who smokes must reconcile their desire for a long healthy life with the fact that smoking is likely to cause lung cancer and other health issues. People have a strong desire to reduce dissonance by altering their views and beliefs or changing their behavior. The smoker might therefore deny the evidence of health concerns associated with smoking or they may attempt to quit smoking.

In a network application, when an actor has positive affinity for two alters who dislike each other, there is dissonance. He must reconcile his positive affinity with his "friends" to their negative affinity toward the third alter. This may cause the actor to change his views toward one of the other alters to reduce the dissonance. Consider the eight possible triads in Figure 4.2 given only positive and negative relationships. You can think of these as "love-hate" triangles. The top four triangles with an even number of negative ties are balanced. The bottom four triangles with an odd number of negative ties have cognitive dissonance. In these four cases there will exist a social force placing pressure on one of the unbalanced triads to move toward balance.

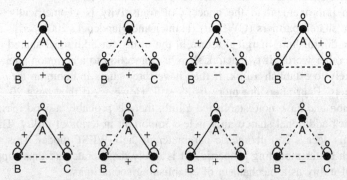

FIGURE 4.2   Eight possible signed triads.

Check on Learning

Note the appropriate social force next to the explanation for the following statements (choose from social forces: Homophily, Reciprocity, Proximity, Prestige, Social Conformity, Transitivity, Balance).

1. Geoff and Ted are friends, but Ted's wife does not like Geoff, so Ted no longer sees Geoff socially.
2. Two individuals belong to the same tennis club.
3. A group of five people who work in the same physical office area trust each other.
4. Joe purchased an iPhone so Bill also purchased an iPhone.
5. Three agents of the same religion spend private time on family outings together.
6. Jake took up golf because the rest of his peers play golf.
7. Mary was invited to dinner by Casey when she took up a new position in the department, Mary returned the invitation and now they share recipes.
8. Lance and Fred are both engineers but are situated at different locations. Despite their separate physical locations they frequently use Skype and emails to share knowledge and ask advice.
9. Charles is respected because he has a degree in law and is very knowledgeable about legal matters.
10. Gillian met Susan at John's party, then again at a local fund raising event managed by John. They got talking and realized they had common interests.

Answers

1. Balance
2. Homophily
3. Proximity
4. Social conformity, Prestige
5. Homophily
6. Social conformity
7. Reciprocity
8. Proximity, Homophily
9. Prestige
10. Transitivity

4.4 GRAPH STRUCTURE

Several theories have natural extensions to graph level measures and properties. Balance, in particular, lays the foundation for Structural Balance and Clusterability in networks that are briefly presented here.

4.4.1 Structural Balance

Structural Balance is a graph level measure that will lie between 0 and 1, just like centralization. A structural balance of 1 means a graph is perfectly balanced, and a 0 means none of the triads are balanced. Figure 4.3a and b is an example of unbalanced and balanced graphs. Each link in the two graphs is labeled as either positive or negative. Balanced triads will reflect one of the four structures in the top row of the signed triads in Figure 4.2 and unbalanced triads will demonstrate a structure in the lower triad row.

4.4.2 Clusterability

Clusterability (Davis, 1967) is an aggregate level measure that describes how well the network can be partitioned into teams. Imagine a parliament of legislative body with two political parties. Suppose that one party's members all like each other, but dislike all members of the opposing party. The other party is similar with all its members liking each other and disliking all members of the opposing party. In this case, we would see a signed adjacency matrix that looks like this:

$$
\mathbf{A} = \begin{pmatrix}
0 & + & + & 0 & - & - \\
+ & 0 & + & - & 0 & - \\
+ & + & 0 & - & - & 0 \\
0 & - & - & 0 & + & + \\
- & 0 & - & + & 0 & + \\
- & - & 0 & + & + & 0
\end{pmatrix}
$$

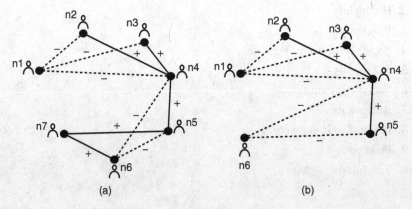

(a)   (b)

FIGURE 4.3   Two signed graphs: (a) unbalanced network and (b) balanced network.

Check on Learning

Draw the network depicted by the above signed adjacency matrix, A, and determine if any clusters are present.

Answer

Clusterability can be determined by reordering the nodes in the adjacency matrix to place positive and negative relations next to each other as shown in the adjacency matrix above. The first three nodes in the adjacency matrix have positive relations in the top left partition of the adjacency matrix, which shows links to other nodes in this group. There are negative ties to other nodes in the network. The lower right partition also has all positive links. This example network is therefore perfectly clustered with three nodes in each group. If there was any dissent within the parties, the graph is no longer perfectly clusterable. Unlike structural balance, clusterability is NOT a sliding scale measure. A graph is either clusterable or it is not.

4.5   AGENT OPTIMIZATION STRATEGIES IN NETWORKS

Some actors in a network will employ various strategies to optimize the value they derive from a social network. Value can be defined in several ways. It may consist of the access to resources that they gain. It may be the power they derive from brokering resources. It may be the social acceptance value they derive from being central. There are several key strategies, which include structural holes, social capital, and link optimization.

4.5.1   Structural Holes

An actor may attempt to maintain a position as the sole broker between two groups. Ron Burt (1995, 2009) describes the emergent network pattern as *structural holes*. Although the actor's position does provide him an informal power in the network, it can be very difficult to maintain. Most individuals view this type of position negatively and these actors can be sanctioned by the group in other ways. Others in the group will inevitably seek to develop redundant ties separate from the broker to reduce their dependence on him and increase their own utility in the network. Actors attempting to maintain structural holes, therefore, must spend time preventing links from emerging and this behavior reduces collaboration, knowledge, and resource sharing across the organization. Conversely, transitivity increases collaboration, knowledge, and resource sharing and can still provide the actor with informal power derived from his ability to connect to others.

4.5.2   Social Capital

The ability of an actor to connect to others in the network and provide access to knowledge and resources through their social connections is known as *social capital* (Bourdieu 1979, 1980, 2008; Coleman, 1988). There is value in collaboration and knowledge. Some managers do not understand this concept and actually unwittingly destroy social capital within their organizations. These managers see their employees chatting in the break-room, going to the designated smoking area to indulge their habit, or otherwise socializing instead of completing some assigned task. The managers view this type of interaction as the employee wasting time or having a poor work ethic. In reality, these social interactions provide unique opportunities for people in disparate parts of an organization to interact and exchange knowledge and information. These activities enable employees to establish new links throughout the organization. These employees now act as high betweenness nodes brokering knowledge and resources. Successful managers will often attempt to create unique opportunities to facilitate social interaction and collaboration and view chatting in the break-room as a valuable investment in organizational capacity and social capital.

Social capital does not come free. There is a cost in terms of time, social expectations, and resources to develop and maintain friendships. Not all actors within an organization will provide the same value, so social capital investments must be made wisely. Link optimization can facilitate the efficient acquisition of social capital.

4.5.3   Link Optimization

Link Optimization is a concept proposed by Cornell sociologist Matthew Brashears (2011). He describes two aspects of social relations: bandwidth and latency, which is referred to in this text as frequency and intimacy, respectively. Frequent positive interactions with an individual will increase the perceived utility of the connection. A friend or coworker that you see several times a week will usually feel like a stronger, more valuable connection than someone you have not seen in years, all else being equal. Intimacy also increases the perceived utility of the connection. As people share secrets, experiences or exchange favors, their level of intimacy increases, but this intimacy does not necessarily entail high frequency of interaction.

Consider two potential actors, a household nanny and your best friend since childhood. The nanny has frequent interaction with you, whereas you might not have seen your childhood friend in years and only speak on the phone a couple times per month. Although the nanny has some level of intimacy in that she works in your home and deals with your children, she has probably not shared the depth of intimate experiences with you as your childhood friend. If you had an important decision to make or needed a favor, who would you feel more comfortable calling upon? On the other hand, if you wanted to know about what happened on last night's episode of your favorite television program, who would you be more likely to ask?

The two aspects of social relations can be plotted as shown in Figure 4.4. The x-axis shows the frequency of interaction and the y-axis shows the intimacy. Different categories of social connections can be expressed in different quadrants

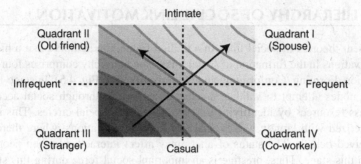

FIGURE 4.4   Pareto optimality space.

of the chart. Quadrant I may include close family members or a spouse. Quadrant II would consist of old, yet close friends or family. Quadrant III consists of strangers. Quadrant IV is made up of coworkers or people with membership in civic organizations. Pareto optimality curves can be expressed over the space represented in Figure 4.4. Relationship strength follows the gradient moving from quadrant III to quadrant I. As the strength of the link increases, the derivative benefit or utility also increases. Unfortunately, the cost associated with maintaining the link may also increase. Connections in quadrant I will have additional expectations associated with maintaining a friendship. In addition to the time cost, they may be called upon for favors and be required to tolerate inconvenience for the sake of maintaining the relationship.

Orthogonal to the strength of relationship is the latency of relationship. Alters classified in quadrant II do not require much investment of time and resources to maintain the association; however, they may still be called upon when needed. This may drive individuals to invest in intimacy to facilitate latent associations as well as develop strong connections. Recently, social media has significantly enabled latency by lowering the cost of maintaining links. This enables people to keep alters in a place where they can be called upon for their utility value when needed with little to no maintenance cost.

In some cases, online media can significantly reduce the costs to maintain a link. Without the cognitive and temporal limitations on the number of links a node can have, some nodes can amass very large numbers of connections. This is common for Web site connections, where once a link is established, there is no cost required to maintain the link. Popular Web sites are able to exploit their prestige and position in the network to acquire more and more links. This enables scale-free networks where there exist hub nodes with many links, although most nodes have relatively few links. The term scale-free is derived from the power-law distribution of node degree, which does not have a scale parameter in the distribution. This type of network structure is not possible in face-to-face networks where cognitive limitations on degree exist. People are not unaware of this phenomenon. A card in the mail is often more meaningful than a "happy birthday" on Facebook.

4.6   HIERARCHY OF SOCIAL LINK MOTIVATION

To present theories of social links in a unified frame, the authors posit a hierarchy of motivations in the formation of social links. The hierarchy comprises four stages: Affiliation, Forming, Conforming, and Optimization (see Fig. 4.5).,In stage 1 (Affiliation), nodes attempt to validate a sense of self-worth through social acceptance or access resources by identifying and affiliating with social circles. This stage is characterized with largely weak social ties to loose affiliations, where there exists a perceived benefit. The status of alters may affect interactions when people first meet in this stage. Thus, prestige is an important social force during this stage. In stage 2 (Forming), nodes attempt to establish links with alters that present the greatest probability of success in terms of validation or resources. Social forces such as homophily, reciprocity, and proximity are paramount. As an actor establishes themselves in the social group, they eventually move to stage 3 (Conforming). In this stage, actors exhibit characteristics of social conformity as they attempt to gain greater group acceptance and establish relationships with high prestige alters. Social forces such as prestige become primary. With an adequate level of group acceptance, a node moves to stage 4 (Optimization). In stage 4, nodes attempt to maintain or increase their utility derived from relationships while minimizing their effort and social costs to maintain links that have been established. Concepts such as transitivity, structural holes, social capital, link optimization, balance, and clusterability provide more significant motivators for link formation.

During situations when small groups are forming and all actors enter a social circle at the same time, there are no established opinion leaders, group norms, or high prestige nodes. This affects the dynamics of link formation. In this situation, stage 1 has already occurred by the assignment of individuals to the group. For example, a cohort of graduate students admitted to a masters degree program have all applied for and been accepted into a program of study. The individuals passed through stage 1 Affiliation in their application process. They will have decided that

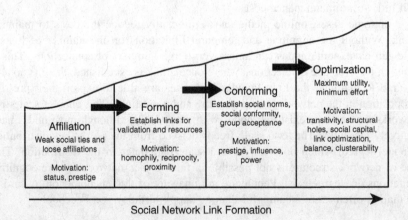

FIGURE 4.5   Four stages of link formation.

TABLE 4.1   Stages of Social Link Motivation and Stages of Small-Group Development

Social link motivation	Small-group development
Stage 1—Affiliation	
Stage 2—Forming	Stage 1—Forming: testing and dependence
Stage 3—Conforming	Stage 2—Storming: intra-group conflict
Stage 4—Optimization	Stage 3—Norming: development of group cohesion
	Stage 4—Performing: functional role relatedness

the graduate program is a worthy pursuit that provides them some sense of self-worth or increased status by hopeful attainment of the degree. Thus the small-group formation begins with nodes in stage 2 Forming. Tuckman (1965) and Tuckman and Jensen (1977) describes this as the first stage in small-group formation, which he labels as "testing and dependence." Individuals determine who they will study with, who they will ask for peer assistance on assignments, and who they will seek social relationships with. With no established authority figures, individuals moving into stage 3 conforming have difficulty in determining what social norms to conform to and whom the group sees as prestigious. This manifests itself with Tuckman's "intra-group conflict." Individuals attempt to assert roles and group values that further their own objectives. As individuals progress into stage 4 optimization, they establish their role in the newly formed group as well as the relationships that maximize their perceived value. Through this process, the group will establish social norms and individuals who epitomize those norms will increase in prestige. Tuckman refers to this as "development of group cohesion." This is where the group establishes a group identity and established routines. At this point, Tuckman introduces an additional stage for small-group development, which he labels "functional role relatedness." In this stage, the group reaches its highest level of productivity. Tuckman combines group formation with task activity and summarizes the combined model as "forming, storming, norming, and performing." Table 4.1 shows stages of social link motivation with corresponding stages of small-group development. In this newly formed group, as new individuals join, group norms and functions are already established. New actors will still test relationships and establish dependence through homphily, reciprocity, and proximity. However, as individuals move toward the subsequent stage, they will have greater tendency to conform to established group norms and there is less intra-group conflict. As these actors optimize their relationships, they assume a new role in the group identity and aid the group in fulfilling its functional role.

4.7   SUMMARY

The following are the main points covered in this chapter

- Social forces of homophily, reciprocity, proximity, prestige, transitivity, and balance drive the formation of social links. Homophily posits that people form relations with those who are similar to themselves, that is, similar beliefs or common interests. Reciprocity occurs where links are reciprocated by another. Proximity by physical location or organizational activity will encourage links where the distance between individuals is minimal. Prestige is the value an individual gains by taking on group norms (social conformity) and values resulting in greater influence within the group. Transitivity purports that if there is a link between actors A and B, and a link between actors B and C, then there is a tendency for a link to develop between actors A and C. Balance suggests that the enemy of my enemy is my friend. Dissonance will occur where one individual has positive links with two alters who dislike each other.

- Structural holes develop where an actor positions him or herself between two groups and actively maintains control as sole broker. Members of the group may view this negatively and develop redundant links to reduce their dependence on the broker and increase their value and effectiveness in the network.

- Social capital is the value an actor obtains from ties with others, providing access to knowledge and resources not previously available. Increasing social interactions and collaboration increases social capital.

- Frequency and intimacy of connections will strengthen relationships, thus optimizing links. Social link motivation comprises four stages:

    — Affiliation: where agents strive to make links that confirm their self-worth.

    — Forming: agents establish links that provide success in terms of validation and/or resources.

    — Conforming: agents adopt social conformity for greater group acceptance.

    — Optimization: agents maximize utility gained from links while minimizing the effort and social costs to maintain those links.

EXERCISES

4.1. Use the three concepts of the "looking glass self" to describe your sense of self-identity.

4.2. How many social circles do you affiliate with? List them and provide a short description of the group. What value do you gain from those relationships?

4.3. Construct a Blau-space. Think of two social circles and two attributes of individuals that might be in those social circles. Draw x and y axes corresponding to the two attributes and plot individuals from the groups based on your numerical evaluation of their attributes. Draw circles that encompass each of the two groups. Do they overlap? What conclusions can you draw about relationship formation?

4.4. How does proximity drive the formation of social relations? How can a common employee lounge area in an office increase collaboration? If you were a manager of a knowledge intensive organization, what kind of policies and practices would you implement to increase social interaction?

4.5. What strategies can an individual employ to increase their prestige within a social group?

4.6. Balance theory applied to a two-party political system would suggest that all members of one political party have affinity for each other and animosity for the opposing party. Why is this not observed in real networks? Can you think of any examples of real-world clusterable networks?

4.7. Consider the last time you entered a new social setting. Perhaps it was your arrival at college, starting a new job, or attending church for the first time. What characteristics did you look for in those individuals whom you initiated relationships with? Why did you develop the relationships you did?

REFERENCES

Asch, S. E. (1955). Opinions and social pressure. *Nature*, **176**:1009–1011.

Asch, S. E. (1956). Studies of independence and conformity: I. A minority of one against a unanimous majority. *Psychological Monographs: General and Applied*, **70**(9):1–70.

Blau, P. M. (1977). *Inequality and Heterogeneity: A Primitive Theory of Social Structure*. Free Press.

Blau, P. M. (1994). *Structural Contexts of Opportunities Structural Contexts of Opportunities*. University of Chicago Press.

Bourdieu, P. (1979). The three states of cultural capital. *Proceedings of the Social Science Research*, **30**(30):3–6.

Bourdieu, P. (1980). The social capital. *Proceedings of the Social Science Research*, **31**(31):2–3.

Bourdieu, P. (2008). The forms of capital. In Biggart, N.W., editors. *Readings in Economic Sociology*. Blackwell Publishers Ltd.

Brashears, M. (2011). Homophily and link atrophy. In Quantification, Metrics and Modeling, Durham, NC. Duke University Workshop on Weak Social Ties.

Burt, R. (1995). *Structural Holes: The Social Structure of Competition*. Harvard University Press.

Cartwright, D. and Harary, F. (1956). Structural balance: a generalization of heider's theory. *Psychological Review*, **63**(5):277–293.

Cheng, J., Romero, D., Meeder, B., and Kleinberg, J. (2011). Predicting reciprocity in social networks. In Privacy, Security, Risk and Trust (PASSAT), 2011 *IEEE Third International Conference on Social Computing (SocialCom)*, pp. 49–56, October.

Coleman, J. S. (1988). Social capital in the creation of human capital. *The American Journal of Sociology*, **94**:95–120.

Cooley, C. H. (1902). Human nature and the social order. In *Scibner's Conference*, New York, pp. 183–184.

Davis, J. (1967). Clustering and structural balance in graphs. *Human Relations*, **20**(2):181–187.

Dunbar, R. (2010). *How Many Friends Does One Person Need?: Dunbar's Number and Other Evolutionary Quirks*. Faber and Faber.

Emerson, R. (1962). Power-dependence relations. *American Sociological Review*, **27**(1):31–41.

Emerson, R. (1969). Operant psychology and exchange theory. In Robert, B. and Bushell, D., editors, *Bevahiorial Sociology*, pp. 379–408. Columbia University Press, New York.

Erikson, E. (1956). The problem of ego identity. *Journal of the American Psychoanalytic Association*, **4**(1):56–121.

Feld, S. (1981). The focused organization of social ties. *American Journal of Sociology*, **86**(5):1015–1035.

Festinger, L. (1957). *A Theory of Cognitive Dissonance*. Stanford University Press.

Friedkin, N. (2001). Norm formation in social influence networks. *Social Networks*, **23**(3):167–189.

Heider, F. (1946). Attitudes and cognitive organization. *The Journal of Psychology: Interdisciplinary and Applied*, **21**(1):107–112.

Homans, G. (1958). Social behavior as exchange. *The American Journal of Sociology*, **63**(6):597–606.

Hunter, D. R. and Handcock, M. S. (2006). Inference in curved exponential family models for networks. *Journal of Computational and Graphical Statistics*, **15**(3).

Johnson, J., Boster, J., and Palinkas, L. (2003). Social roles and the evolution of networks in extreme and isolated environments. *Journal of Mathematical Sociology*, **27**:89–121.

Kadushin, C. (2012). *Understanding Social Networks: Theories, Concepts, and Findings*. Oxford University Press, illustrated edition.

Killworth, P. and Bernard, H. (1979). Informant accuracy in social network data iii. *Social Networks*, **2**:19–46.

Lospinoso, J., McCulloh, I., and Carley, K. (2009). Utility seeking in complex social systems: an applied longitudinal network study on command and control. In *Social Networks and Multi-Agent Systems Symposium (SNAMAS-09)*, Edinburgh, Scotland. Heriot-Watt University, SSAISB.

Marcia, J. E. (1966). Development and validation of ego identity status. *Journal of Personality and Social Psychology*, **3**:551–558.

Mayhew, B. H. and Levinger, R. L. (1976a). On the emergence of oligarchy in human interaction. *American Journal of Sociology*, **81**(5):1017–1049.

Mayhew, B. H. and Levinger, R. L. (1976b). Size and the density of interaction in human aggregates. *American Journal of Sociology*, **82**(1):86–110.

Mcculloh, I., Geraci, J., Turner, J., and Matthews, M. (2010). Network centrality and PTSD. In *Proceedings of SunBelt XXX, Social Networks and Health. International Network for Social Network Analysis*.

McCulloh, I. A. (2013). Social conformity in networks. *Connections*.

McPherson, M. (1983). An ecology of affiliation. *American Sociological Review*, **48**(4):519–532.

McPherson, M., Lovin, L., and Cook, J. (2001). Birds of a feather: homophily in social networks. *Annual Review of Sociology*, **27**(1):415–444.

Milgram, S. (1963). Behavioral study of obedience. *Journal of Abnormal Psychology*, **67**:371–378.

Milgram, S. (1974). *Obedience to Authority: An Experimental View*. Harpercollins.

Molm, L. (1990). Structure, action, and outcomes: the dynamics of power in social exchange. *American Sociological Review*, **55**(3):427–447.

Molm, L., Takahashi, N., and Peterson, G. (2000). Risk and trust in social exchange: an experimental test of a classical proposition. *American Journal of Sociology*, **105**(5):1396–1427.

Ridgeway, C. L. (1978). Conformity, group-oriented motivation, and status attainment in small groups. *Social Psychology*, **41**(3):175–188.

Ridgeway, C. L. (1982). Status in groups: The importance of motivation. *American Sociological Review*, **47**(1):76–88.

Rogers, C. R. (1961). *On Becoming A Person*. Houghton Mifflin.

Sailer, K. and McCulloh, I. (2011). Social networks and spatial configuration–how office layouts drive social interaction. *Social Networks*.

Tuckman, B. W. (1965). Developmental sequence in small groups. *Psychological Bulletin*, **63**(6):384–399.

Tuckman, B. W. and Jensen, M. (1977). Stages of small-group development revisited. *Group & Organization Management*, **2**(4):419–427.

Yeung, K.-T. and Martin, J. L. (2003). The looking glass self: an empirical test and elaboration. *Social Forces*, **81**(3):843–879.

Zafirovski, M. (2005). Social exchange theory under scrutiny: a positive critique of its economic-behaviorist formulations. *Electronic Journal of Sociology*, **2**:1–40.

SUBGROUP ANALYSIS

We must remember that one determined person can make a significant difference, and that a small group of determined people can change the course of history.

—*Sonya Johnson*

Learning Objectives

1. Know the difference between Newman, Consecutive Correlation (CONCOR), and user-defined grouping algorithms.
2. Develop a blockmodel of a social network based on one of the grouping algorithms.
3. Demonstrate the ability to create and interpret a hierarchical clustering diagram, density matrix, reduced network, and attribute analysis table.

5.1 SUBGROUPS

Subgroup analysis is an important approach for understanding the structure of a network as well as assessing organizational risk in networks. Subgroup analysis is carried out to more fully understand the clustering of nodes into subgroups, usually linked by a common factor or positioned close to each other within the network (Wasserman and Faust, 1994; Krackhardt, 1994; Kadushin, 1995). A substantial amount of prior research has been carried out on the groupings of actors who interact frequently and on how individuals are influenced by these interactions. Network groups can have sparse or dense interactions within the group. Early organizational studies reported that large organizations comprise separate subgroups showing dense interactions (Freeman, 1992). Subgroups within organizations were found to link in order for the organization to evolve or for reasons of management efficiency (Simon, 1962; Granovetter, 1973). Organizations of today are focused on teams and group connectivity and can affect the efficiency and effectiveness of work practices, competitive advantage, and ultimately organizational survival. Interactions between groups impact not only strategic alignment and performance

Social Network Analysis: with Applications, First Edition.
Ian A. McCulloh, Helen L. Armstrong, and Anthony N. Johnson.
© 2013 John Wiley & Sons, Inc. Published 2013 by John Wiley & Sons, Inc.

in the attainment of organizational goals but also the creation and sharing of knowledge that in turn feeds innovation. Analysis of the interactions between group members and external nodes is also of interest when measuring the isolation of the group. In some circumstances isolation may be desirable, but undesirable in others.

In this chapter, we must first paint the context for which subgroup analysis is useful. An overview of organizational theory is presented as a foundation to aid our understanding of groups within organizational contexts. The structural importance of subgroups is then discussed including derivation of the probability of groups forming at random. With this context in place, three popular subgroup heuristics are presented and compared: Newman–Girvan (2004), CONCOR (Breiger et al., 1975), and user-defined grouping approaches. Common methods for analyzing the groups are demonstrated on the Newman–Girvan grouping heuristic.

5.2   ORGANIZATIONAL THEORY

Organizational theory is a collection of sociological and psychological theories of groups for the purpose of improving productivity, maximizing efficiency, and facilitation of group problem solving. There are three common perspectives for understanding organizational theory: classical, neo-classical, and environmental. The *classical perspective* is concerned with increasing efficiency and productivity by designing organizational structures, rules, and expertise. An early pioneer in the study of organizations, Frederick Taylor (1911), introduced *scientific management* for "knowing exactly what you want men to do and then see in that they do it in the best and cheapest way." He identified four principles: (i) scientific measurement of performance, (ii) training of workers, (iii) cooperation between management and workers to ensure productivity, and (iv) division of labor where workers and management specialize in specific tasks and roles.

Another classical theory is *Bureaucracy theory*, introduced by Max Weber (Gerth and Wright Mills, 1948), which focuses on the part of an organization that implements rules, laws, and functions of the group. Individuals assume well-defined roles in this organization with established policies for conduct and performance. Neutral parties, for example, make decisions for career advancement based on individual performance, thus removing personal bias from the process.

There are several important limitations for viewing organizations from the classical perspective. Most important is Taylor's initial assumption . . . "knowing exactly what you want men to do . . ." Scientific management and bureaucracy are better suited to production and manufacturing processes or standard administrative functions, where the task is well defined and repetitive. For knowledge-intensive organizations focused on creative problem solving or knowledge development, there is no concept of "what you want men to do." It is by definition of purpose left undefined. Furthermore, the classic approach omits human needs and concerns. Human error, the variability of performance, motivations, and individual goals are rarely considered with this approach. These limitations provide the basis for a neo-classical approach.

The most fundamental principle in the *neo-classical perspective* is the *Hawthorne effect* (Mayo, 1933, 1945; Whitehead, 1938; Roethlisberger and Dickson, 1939; Roethlisberger, 1941), which emerged out of a series of human productivity experiments at Western Electrics Hawthorne Works facility outside of Chicago, IL. The Hawthorne effect is a phenomenon where individuals alter their behavior as a result of their awareness that they are being observed and measured. This introduced a social-psychological dimension to workplace productivity. Mayo's work in particular asserted that humans were unique and motivated differently and are not interchangeable parts. His work has been criticized as counter to Taylor's scientific management approach. However controversial, Mayo paved the way for organizational theorists to study motivation, employee satisfaction, peer social relations, and leadership in organizations.

Within the context of social link theory (Chapter 4), motivation, employee satisfaction, and peer relations can perhaps be better understood. An individual worker will initially affiliate with an organization in order to access financial or other resources of benefit. The nature of this organization becomes a social circle with an inherent group culture. Individuals are driven to form social relations to gain acceptance. Individuals conform to the organizational culture to achieve validation and prestige within the culturally defined norms. These standards might be performance driven, qualification based, or derived from interpersonal relations. Financial incentives might provide a sense of value in certain organizations, such as financial investment firms or large corporations. We argue that it is not so much the money as it is the value it conveys that is important. In the US military, status is derived from combat tours and specialized military qualifications such as ranger, or paratrooper. Financial incentives are not as effective as status in motivating this workforce. Other extensions of the Hawthorne studies have revealed non-financial incentives for productivity. Thus, an understanding of social link theory provides an alternate frame of reference to understand group and individual motivation. This is at the heart of informal power brokerage. One may ask why a low ranking individual, who is highly central within an organization's social network, may have high informal power, which is more influential in motivating group behavior than the formal power a designated leader is assigned. We contend that this is precisely the issue that Mayo was concerned with understanding, and social network analysis provides an insightful frame of reference to conceptualize this phenomenon.

The most recent perspective in organizational theory is the *environmental perspective*. This approach argues that there is no universally best practice for organizing, leading, or decision making for a particular type of organization. Organizational design must therefore not only consider the function of the organization, but it must also contend with internal social dynamics of the individual members, current and desired organizational culture, informal power brokers and opinion leaders, changing external environment that may affect the organizations purpose, and more. This approach requires a manager to understand the social terrain of their employees and recognize that the organization may not actually be interacting as programmed.

Regardless of perspective, organizations can be classified into one of six structure types (Daft and Armstrong, 2009): prebureaucracy, bureaucracy, postbureaucracy, functional, divisional, and matrix. The *prebureaucracy* structure lacks any standardization of role or task and a central leader makes all decisions. This type of structure is usually confined to small organizations engaged in completing simple tasks.

The *bureaucracy* is defined as an organization that has standardized roles and is focused on efficiency. Weber proposes three criteria for a bureaucracy: (i) clearly defined roles and responsibilities, (ii) hierarchical structure, and (iii) value for merit (as defined by the organizational culture). This structure is well suited for larger organizations engaged in well-defined activities. Again, the primary advantage of the bureaucracy is efficiency; however, it is usually achieved at the cost of reduced flexibility and innovation.

The *postbureaucracy* (Heckscher and Donnellon, 1994), also referred to as *organic*, is the opposite of a bureaucracy. There are no standardized roles, responsibilities, or formal leadership. There is no hierarchy. Decisions are made through discourse and consensus. This type of organization is common among cooperatives, community organizations, and nonprofit organizations. It encourages participation and empowers individuals who may not otherwise have a voice in the decision-making process. The primary advantage of this organization is its flexibility, innovation, and commitment of membership; however, these goals are achieved at the cost of efficiency.

The *functional structure* is defined as an organization where groups are formed based on specialized expertise. For example, there may exist a sales team, an engineering division, and an accounting office. Or in a manufacturing process, casting, milling, shaping, joining, and assembly might all fall in separate divisions, where production flow moves back and forth between divisions and competes with other product lines for processing. Functional structures increase skill capacity through divisional mentorship, training, and oversight. However, they lose some efficiency in cross-divisional collaboration that might inhibit the development of new products and cross-disciplinary problem solving. This structure is best suited for organizations that conduct a limited set of tasks or produce a limited amount of products at high volume and low cost.

The *divisional structure* is constructed such that it has each functional element within it and can operate almost independently from other divisions. Sometimes divisions are organized geographically such as the United States, Australia, or EU divisions. Alternately, divisions may be organized around product lines such as consumer goods, commercial goods, and industrial goods. Each division may even have its own marketing and sales departments. The advantage of this structure is its flexibility to adjust to changing environments and operate independently; however, it comes at the cost of reduced functional expertise.

The *matrix structure* combines the functional and divisional structures. The organization frequently uses teams to accomplish tasks. Matrix structures may have *weak/functional* management, where a project manager oversees the product, whereas functional managers retain most decision making and control over the functional employees. Other organizations may have a *strong/project* management, where project managers control the project and functional leaders provide technical expertise and assign resources as needed. *Balanced* management shares responsibility equally between functional managers and project managers. This organizational structure strikes a hybrid between other structures with the associated strengths and weaknesses.

Depending on the organization's purpose and function, subgroup analysis provides a valuable frame of reference for understanding organizations. Social links may be defined between individuals as affinity, advice seeking, communication, collaboration, perceived usefulness, frequency of interaction, or other forms of relationship. Identifying subgroups by defined organizational structure allows the network analyst to evaluate whether the organizational structure is performing as designed or if a new structure might improve performance. Additionally, methods such as Newman grouping can identify self-organizing subgroups that may provide insight into organizational design.

The purpose of a manager or leader in any organization is to define goals and motivate organization members to maximize their productivity and efficiency. Social network analysis provides powerful tools and theory to understand important social dynamics that affect every aspect of an organization. The remainder of this chapter will present several techniques and theories for understanding subgroups within social networks. Identifying subgroups also allows an investigator to study networks of organizations in addition to individual actors.

5.3   RANDOM GROUPS

Consider for a moment that the probability of a node being connected to another node occurs completely at random with the flip of a coin (probability is 0.5). Then, the probability of getting a unique sequence of 1s and 0s is $(\frac{1}{2})^{(n-1)}$. Therefore the probability of two nodes being exactly structurally equivalent (have exactly the same connections) is also $(\frac{1}{2})^{(n-1)}$. In an eight-node network for example, the probability of two nodes being structurally equivalent is $(\frac{1}{2})^{(8-1)} = 1/128 = 0.0078$. If we looked at a network of only 30 nodes this probability would be 0.0000000186. Therefore, even in small networks the probability of two nodes even knowing similar people is quite small. However, in practice it is not unusual for people to have similar acquaintances. This is because there is some reason for the social interactions, such as shared membership in an organization or some of the social forces presented in Chapter 4. The reason for structural grouping in networks is precisely what makes subgroup analysis so important.

5.4   HEURISTICS FOR SUBGROUP IDENTIFICATION

Networks can be partitioned into subgroups based on the similarity between nodes. There are several ways a node can be similar, however, which has led to multiple approaches for assigning group membership. Three approaches are presented in this chapter: Attribute Defined, CONCOR (Breiger et al., 1975), and Newman–Girvan (2004).

5.4.1 Attribute Defined

The simplest approach to assign nodes to subgroups is by a predefined attribute. This is often done in organizational studies where the attribute is the work section within an organization. For example, a company might have sections such as shipping and receiving, production, sales, customer service, research and development, and management. There is usually a desire to understand how collaborative these sections are with one another. One might hope that there is a lot of interaction between customer service and research and development or between production and sales for example. Assigning individuals by a defined attribute allows exploration of intergroup collaboration.

Assigning nodes to groups based on a predetermined attribute is essentially a straightforward method to defining the groups without the use of any algorithm or heuristic approach. The attributes may represent node characteristics to study potential homophilous social circles, also known as *Blau-space*. Alternatively, the attribute might be the pre-designated grouping such as formal structures within an organization. When groups are pre-designated, subgroup analysis methods such as block models can be used to explore organizational collaboration and function.

5.4.2 Consecutive Correlation (CONCOR)

Consider an adjacency matrix for a social network. The CONCOR heuristic correlates the rows or columns of the adjacency matrix, creating a new matrix that contains the correlation between node connections. CONCOR then takes iterative correlations of this new matrix. This process will converge such that nodes are correlated with a +1 or a −1, thus partitioning the network into two groups. This procedure can be executed multiple times, each time bifurcating the network in two. Therefore, executing the algorithm three times should partition the network into eight groups. It is possible, however, that no nodes would be assigned to one of the groups in the case of structural equivalence between nodes.

CONCOR is typically used to identify role redundancy within a network. There is a certain amount of redundancy that is necessary within an organization. Identifying nodes that are similar in their connections can assist the analyst in strengthening and coordinating activities of redundant nodes. In other cases, there may be too much redundancy, such as after a merger, acquisition, or joint venture. Management may wish to reassign individuals to other activities within the organization or make other labor decisions. CONCOR can assist by identifying similar individuals.

5.4.3 Newman–Girvan Grouping

Another approach is the Newman–Girvan grouping, commonly referred to as simply Newman grouping. The Newman grouping iteratively identifies edges that are high in betweenness. This is similar to betweenness centrality; however, the algorithm looks for the edges that fall on geodesics most frequently. The highest edge in betweenness is removed and the edge betweenness is recalculated. The next edge that is highest in betweenness is removed and so forth. This process is repeated until a specified number of subgroups are found or until a specified ratio of in-group links to out-group links is achieved. Thus, the number of subgroups can be specified or the algorithm can determine the number of subgroups.

The Newman group is excellent for identifying network clusters. The clusters are naturally formed groups that essentially have no probability of occurring at random, as shown earlier. An analyst can explore the network to attempt to identify the reason for the social grouping. This is a common approach when there are no attribute defined groups.

Check on Learning

Which subgroup analysis method is used to find clusters of related nodes and/or cliques?

Answer

Newman–Girvan grouping. The attribute defined groupings do not necessarily consist of related nodes. They could be anything that the analyst defines. The CONCOR grouping identifies nodes with similar connections, which usually identifies role redundancy, but not necessarily clusters. The Newman–Girvan grouping iteratively deletes links with high betweenness, revealing clusters of related nodes.

5.5 ANALYSIS METHODS

We examine three analysis methods for investigating subgroups in a network. They are the group membership method, the hierarchical clustering method, and the block model method (White et al., 1976). These methods will be demonstrated using example network graph and adjacency matrix (Fig. 5.1 and Table 5.1).

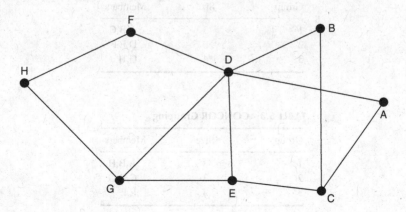

FIGURE 5.1   Example network.

TABLE 5.1   Adjacency Matrix for Example Graph

	A	B	C	D	E	F	G	H
A	0	0	1	1	0	0	0	0
B	0	0	1	1	0	0	0	0
C	1	1	0	0	1	0	0	0
D	1	1	0	0	1	1	1	0
E	0	0	1	1	0	0	1	0
F	0	0	0	1	0	0	0	1
G	0	0	0	1	1	0	0	1
H	0	0	0	0	0	1	1	0

5.5.1   Group Membership

The *group membership* analysis method investigates the assignment of nodes to different groups. It can be seen in Table 5.2 and Table 5.3 that the Newman grouping and CONCOR grouping will group the nodes in the network differently. In the Newman grouping, the groups are close together in the network. Looking once more at the Example Network graph (Fig. 5.1), Group 1 comprising nodes A, B, and C appears at the right of the network. Group 2 consisting of nodes D, E, and F appears in the center of the network. Nodes G and H, forming Group 3, appear at the left of the network. In contrast, the groups in the CONCOR grouping are based on the correlation or similarity of connections. Nodes A, B, and H are all similar in their connection to node D and their lack of connections to nodes E and F. As node C is connected to node E, it is excluded from this group. Nodes C, D, and G are similar in their connections to Group 1; however, none of them are connected to each other. This group may fulfill similar or redundant roles in the

TABLE 5.2   Newman Grouping

Group	Size	Members
1	3	A,B,C
2	3	D,E,F
3	4	G,H

TABLE 5.3   CONCOR Grouping

Group	Size	Members
1	3	A,B,H
2	3	C,D,G
3	4	E,F

organization. Nodes E and F are similar in their connection to Group 2 and their lack of connection to Group 1.

Check on Learning

Which grouping algorithm do we need in order to find out the following information from our social network of employees and clients? There are only two node types in our network: agent (employee) and agent (client). Employee node information includes their name, store location, department, gender, and pay level. Client node information includes their name, address, age, and gender.

1. Does gender matter when linking employees and clients? We want to know which female employees connect with female clients, female employees with male clients, and finally male employees with male clients.
2. We want to remove duplications in the relationships between employees and clients. Which employees are connected to the same clients?
3. We want to know who works in the same department. All employees have links with the other employees in their department but few links with employees in other departments.
4. We want to minimize our pay costs. Which employees present the highest cost in pay and which are the lowest pay cost?
5. Who are the clients who connect with employees in more than one department?

Answer

1. Newman grouping.
2. CONCOR grouping.
3. Attribute grouping on department, and Newman grouping.
4. Attribute grouping on pay level.
5. CONCOR grouping.

5.5.2 Hierarchical Clustering

The *hierarchical clustering diagram* depicted in Figure 5.2 provides a visual chart of the similarity between nodes in a network. The dots along the top row indicate that at some level all of the nodes are isolated. At the next level, nodes G and H have the closest relationship. At the next level, nodes A and C have the closest relationship. At the next level, nodes E and F have the closest relationship. Then, at the next level, node D is added to the E–F group. Then B is added to the A–C group. At this point, the network is clustered into three groups, which is what the Newman grouping returned as the optimal number of subgroups. However, you

```
Hierarchical Clustering Diagram

          B A C D E F G H

  Level   2 1 3 4 5 6 7 8
  -----   - - - - - - - -
      0   . . . . . . . .
 -0.066   . . . . . . XXX
 -0.000   . XXX . . . XXX
 +0.053   . XXX . XXX XXX
 +0.111   . XXX XXXXX XXX
 +0.161   XXXXX XXXXX XXX
 +0.161   XXXXX XXXXXXXXX
 +0.000   XXXXXXXXXXXXXXX
```

FIGURE 5.2   Hierarchical clustering chart produced in ORA for the example network of Figure 5.1.

could continue grouping and connect the D–E–F group with the G–H group and finally connect the whole network as one complete group. This diagram provides some insight into the group formation and the sensitivity of the network to a grouping threshold.

5.5.3   Block Model

Once the groups have been determined, a *block model diagram* depicted in Figure 5.3 provides some very important insight into group dynamics. The nodes are arranged in the rows and columns of the adjacency matrix such that they are listed next to other nodes within their group. Lines are drawn between the groups. Compare the block model (Fig. 5.3) to the original adjacency matrix, Table 5.1. In this case the original adjacency matrix is listed in the same order as the subgroupings, but this is not always the case.

Block densities can be determined by calculating the number of links in each block divided by the number of possible links. For this example, in the upper left block between nodes A,B,C, and themselves (1,2,3) there are $n(n-1)$ or six possible links. Recall that the diagonal of the adjacency matrix is undefined, disregarding the possibility of reflexive links. There are actually four links present

Block Model

```
        1 2 3   4 5 6   7 8                1 2 3   4 5 6   7 8
       -------------------------          -------------------------
 1 A |     1 | 1     |     |        1 A |       |       |       |
 2 B |     1 | 1     |     |        2 B | 0.67  | 0.33  | 0.0   |
 3 C | 1 1   |     1 |     |        3 C |       |       |       |
       -------------------------          -------------------------
 4 D | 1 1   |   1 1 |   1 |        4 D |       |       |       |
 5 E |     1 | 1   1 |     |        5 E | 0.33  | 1.0   | 0.33  |
 6 F |       | 1 1   | 1   |        6 F |       |       |       |
       -------------------------          -------------------------
 7 G |       |     1 |   1 |        7 G |       |       |       |
 8 H |       | 1     | 1   |        8 H | 0.0   | 0.33  | 1.0   |
       -------------------------          -------------------------
```

FIGURE 5.3   Block model chart produced in ORA for the example network of Figure 5.1.

Block Reduced Network

```
      1 2 3
  1   1 0 0
  2   0 1 0
  3   0 0 1
```

Network density: 0.333333

FIGURE 5.4 Block reduced chart produced in ORA for the example network of Figure 5.1.

out of the six possible, so the block density is 0.67. The block connecting A, B, C to D, E, and F (4,5,6) has n^2 or nine possible links, as the diagonal is included in this block and all potential entries are possible. The block density is therefore three out of nine possible or 0.33. The block densities provide insight into the coordination and connections between subgroups in the network. Figure 5.4 depicts the construction of a *reduced network* where the nodes of this new network are the subgroups and the links are connections between subgroups. This is extremely powerful for exploring collaboration within an organization. Subgroups with high informal power and brokerage or diffusion reach can also be determined.

There are three ways to determine the connections between subgroups. The *fully connected* approach would argue that for a subgroup to be connected to itself or another group, all of the nodes of one group must be connected to all of the nodes of the other group. Under this criterion, subgroup 2 is connected to itself because D is connected to E and F and E is connected to F. Subgroup 3 is connected to itself because G and H are connected to each other. Subgroup 1 is not connected to itself because node A is not connected to node B.

The *minimally connected* approach would argue that if one member of one group is connected to one member of another group, there is a connection between the groups. Under this criterion, the only groups that are not connected are Groups 1 and 3, because there are no connections between node G to A, B, or C and there are no connections between node H to A, B, or C.

The most widely accepted method for determining subgroup links in a reduced network is to compare the block density with the original density of the network (input network density). In this example, the original network density is 0.393. The only block densities that exceed the input network density are those along the diagonal. Thus the input network density approach would identify cohesive subgroups.

The *external/internal link analysis* can provide some further insight into how well groups collaborate with each other within the network. This analysis compares the number of internal links within the subgroup to the number of links that nodes in the subgroup have with nodes outside the subgroup. The silo index provides a score that ranges from -1 to 1 to show how isolated a particular subgroup is. If 50% of the links from a subgroup are internal or external, then the silo index is 0. If a subgroup has more internal links than external links the silo index will be positive (Table 5.4).

TABLE 5.4   External/Internal Link Analysis

Group	Internal link count	External link count	Internal links%	Silo index
1	4	6	40	−0.2000
2	6	10	37.50	−0.2500
3	2	4	33.33	−0.3333

Example 5.1

For the network depicted in Figure 5.5 determine the Newman grouping.

The first step of performing a Newman grouping is to label the edges so that they can be treated as nodes for betweenness calculations. Figure 5.6 illustrates the process.

Next, we calculate edge betweenness for the entire network using the methods given in Chapter 2. In this case, edge "f" has the highest betweenness and is therefore removed. See Figure 5.7. At this point it is intuitively clear to see that edge "a" and edge "b" have the same betweenness. We randomly choose "b,", but

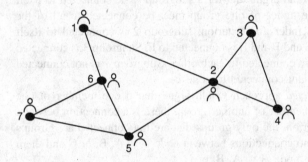

FIGURE 5.5   Example network for Newman grouping.

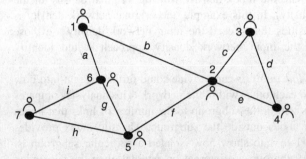

FIGURE 5.6   Network with edge labels.

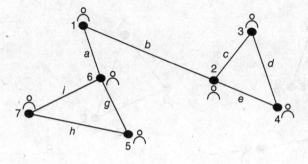

FIGURE   5.7   Highest betweenness edge removed.

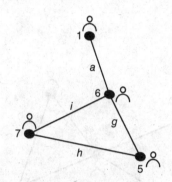

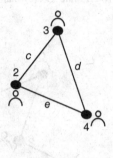

FIGURE 5.8   Newman subgroup emerges.

one might choose "a" as well based on some prior iteration of the algorithm. After removal of edge "b", we see the first fracture of the network into two separate pieces illustrated by Figure 5.8.

Calculating edge betweenness again reveals that edges "i" and "g" have the same betweenness value. Choosing "i" arbitrarily produces the network depicted in Figure 5.9.

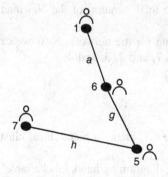

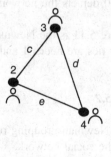

FIGURE 5.9   Iterate of Newman algorithm.

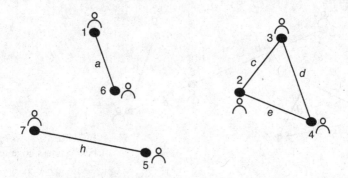

FIGURE 5.10   Iterate of Newman algorithm.

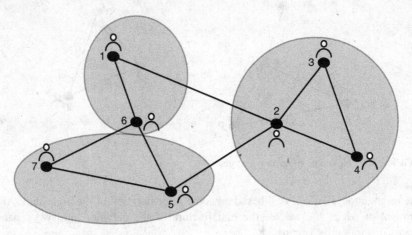

FIGURE 5.11   Newman groups.

The final edge betweenness calculation identifies "g" as the edge to be removed giving us the final version of the network with three clusters identified. Figure 5.10 depicts the network after the final iteration of the Newman grouping algorithm.

Figure 5.11 is the Newman grouping for the network. Now we can see that the closest ties are nodes 1 and 6, 5, and 7, and 2, 3, and 4.

Example 5.2

Using the Newman grouping of Example 5.1, create a hierarchical clustering diagram for the social network.

To create the hierarchical clustering diagram by hand, label a table's columns with the closest ties next to each other, and the rows by the number of steps (or levels) it takes to proceed from a completely connected network to a completely

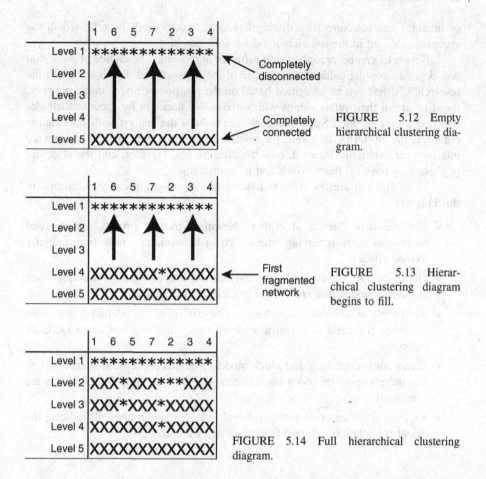

FIGURE 5.12 Empty hierarchical clustering diagram.

FIGURE 5.13 Hierarchical clustering diagram begins to fill.

FIGURE 5.14 Full hierarchical clustering diagram.

disconnected network (Fig. 5.12). Next, taking note of successive iterates of the Newman algorithm, determine the structure of the first nonconnected network. In this case, it appears in Figure 5.8. We note the structure, and our diagram slowly begins to fill with very useful information as to the emergence of subgroups based on structure (Fig. 5.13).

The final form of the hierarchical table shows the progress of the Newman algorithm from a connected network to a fully fragmented network with no ties (Fig. 5.14).

5.6 SUMMARY

Where the network is an organization, the type and structure of the organization will provide the foundation for groupings within the network and the density of interactions between groups. Factors such as the hierarchy (bureaucratic, post-bureaucratic, or organic) and structure (functional, divisional, matrix) will be reflected in the groupings evident in the organizational network graph.

Networks can be grouped using different algorithms; the choice of algorithm will depend upon the nature and structure of the network and the objectives of the research. Clusters can be identified based on the position of nodes in the network, the structure of their relationships with surrounding nodes or by specific attributes that nodes may display. Subgroup analysis provides the analyst with information regarding the clustering of nodes, the existence of cliques, whether the clusters are internally or externally focused, how the clusters link together, and the effect all of these may have on the network and its operations.

The following are the main points relating to subgroup analysis covered in this chapter:

- The Newman–Girvan algorithm (Newman groups) groups nodes based on progressively removing edges high in betweenness, reflecting naturally formed clusters.

- The CONCOR algorithm divides the network into two groups multiple times to identify nodes that are similar in their connections.

- Attributes associated with nodes can be used to define subgroups, removing the need to perform analyses using the Newman and CONCOR algorithms.

- Hierarchical clustering and block model diagrams illustrate in a tabular format the grouping of nodes into clusters and illustrate how these clusters are connected.

- An analysis of external versus internal links for subgroups will indicate the extent of isolation (siloing) from the rest of the network.

CHAPTER 5 LAB EXERCISE

Subgroup Analysis

Learning Objectives

1. Analyze subgroups in a network using the Newman clustering algorithm.
2. Analyze subgroups in a network using the CONCOR clustering algorithm.
3. Analyze subgroups in a network using Attributes.
4. Visualize a network and color nodes by Newman, CONCOR, or Attributes.

Enter the data for the network Subgroups1 into ORA. The tables of data for this network can be found in Table B.8 in Appendix B. Save the meta-network as *Subgroups1.xml*.

Visualize the network and save the network diagram as a .jpg. The *Subgroups1* network should look like as follows:

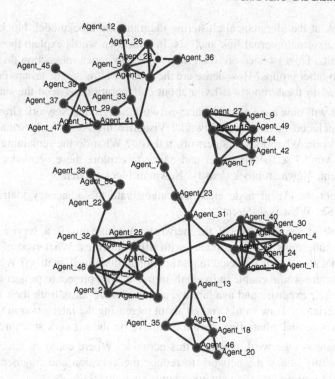

Newman Subgroups

Newman grouping is based on Betweenness. The links between agents within a subgroup will have a purpose, so Newman clustering will show groups of agents with a reason to connect. Generate a Locate Subgroups report, select Newman grouping and ensure the following options are turned on:

- Create group membership network
- Create block diagram/silo analysis
- Create hierarchical clustering diagram

And also at the bottom

- Add a new meta-network with the located groups

Generate the report in HTML format and enter the location for the file and your report filename.

Save your subgroup reports.

This report will not only produce an HTML report, but also a new meta-network showing the groups and how they are linked. Visualize the new meta-network. The Newman groups are shown as the additional pink nodes and links.

Check the hierarchical clustering diagram, blockmodel, and block densities and find the silo index via external/internal link analysis in the Newman report.

1. Look at the hierarchical clustering diagram, the blockmodel, block densities and external/internal link analysis. In your own words explain the silo index results. How isolated are the groups and how much interaction do they have with other groups? How dense are the groups? How are the groups connected? What do these reports tell you about collaboration between the subgroups?

2. You will now have an extra meta-network Subgroups Network Groups. What is included in this meta-network? Visualize the Agent × Newman group's network. What does this network tell you? What do the remaining networks tell you? Use the visualizer and editor to explore these networks: Agent × Agent, Newman block density, Newman block model.

3. Select the Agent node class and investigate the adjacency matrix via the Editor. What is different?

4. Look at the structure of the network. Assume this is a service firm and the subgroups are departments with different roles. What type of organizational structure is reflected (e.g., functional, divisional, matrix)? What type of structure would enable individuals to move from project to project depending on their expertise and availability, and thus share ideas from their broadened experience? How would you go about increasing the interaction in the current structure and what effect would this have on the network structure?

5. Where are the weak points in this network? Where could an adversary successfully attack the network to reduce the cohesion and fragment it? How could you protect the network against such attacks?

Visualize the Subgroups1.xml network and color the nodes by Newman grouping. To do this select Display in the visualizer, Node Appearance, Node Color, Color Nodes by Newman grouping. You will see that the network diagram shows groups in seven different colors.

CONCOR Subgroups

CONCOR grouping is based upon structural equivalence. It will highlight redundancies in the network where nodes have similar connections. Generate the Locate Subgroups report using CONCOR for this meta-network. Select CONCOR and use level 1 for this exercise. The group should bifurcate into two groups. Use the same options as previously:

- Create group membership network
- Create block diagram/silo analysis
- Create hierarchical clustering diagram

And also at the bottom

- Add a new meta-network with the located groups

Look at the results for the two groups. Now run the report again but this time set three levels for this exercise. Your report should produce eight groups; however one or two may not have many entries.

1. Explain the results for the block model and external/internal link analysis.

2. Look at the additional networks generated by the CONCOR analysis. Visualize these and analyze what these tell you.

3. Select the Agent node class and investigate the adjacency matrix via the Editor. What is different?

4. This organization has some obvious redundancy as shown by the report. Highlight those areas where redundancy is occurring by circling them on the network graph so that management can decide how to redeploy agents in those areas being duplicated.

Visualize the Subgroups1.xml network and color the nodes by CONCOR grouping. To do this select Display in the visualizer, Node Appearance, Node Color, Color Nodes by CONCOR grouping. Enter the Number of Splits as 1. The network will be shown in two colors and groups that are structurally equivalent will have the same color. Display it again but this time, enter the Number of Splits as 2. How many different colored CONCOR groups are there? There should be 4.

Grouping by Attribute

You can also group nodes together by attributes. Where Newman and CONCOR analyze groups are based on links, groups can also form based on common attributes.

Add the attribute data from Table B.9 in Appendix B to the Agent network in the *Subgroups1.xml* meta-network. The attribute to be added is the project to which they are currently assigned.

To add an attribute to a node type, select the Agent node class and then the Editor tab. On the right-hand side select Create in the Attributes section. Name the attribute Project and the Type will be Number Category. Select Create, then Close. You can now enter in the data from the above table.

Run the Locate Subgroups report. Use the entire meta-network Subgroups1, and only this network (do not include other networks generated by Newman or CONCOR analysis). Select the Newman, CONCOR, and Attribute grouping algorithms, then choose the attribute Power. Look at the groupings of nodes, the blockmodel and the external/internal link analysis. The nodes in these groups are quite distributed, so the silo index for each group will be a relatively high negative figure, illustrating that the nodes have more links outside their clusters than within. Compare this with the silo index for the Newman and CONCOR groupings.

Once completed, visualize the network and display the node color by the Project attribute. Display, Node Appearance, Node Color, Color Nodes by Attribute or Measure, choose the Project attribute and select the colors you wish for each project number by clicking on the color swatch, or accept the default color assigned. Your network will now color the nodes by the project on which they work.

In this laboratory exercise you have generated Subgroup reports and analyzed their contents in relation to the example network. You have produced subgroup analyses using the Newman and CONCOR algorithms and also grouped by Attributes.

Visualizations of the network produced groups of nodes based upon Newman and CONCOR groups as well as attributes.

EXERCISES

Use the above networks to complete the following exercises.

5.1. Calculate the Newman grouping without the use of technology.

5.2. Create a hierarchical clustering diagram for each network.

5.3. Create the block density matrix for each network.

5.4. Compare your groupings and models for these three networks and briefly explain the differences.

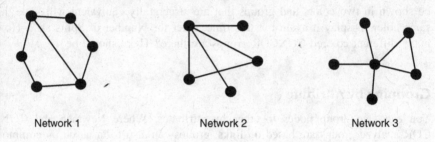

Network 1   Network 2   Network 3

REFERENCES

Daft, R. L. and Armstrong, A. (2009). *Organization Theory and Design*. Nelson Education.

Freeman, L. (1992). The sociological concept of 'group': An empirical test of two models. *American Journal of Sociology*, **98**:55–79.

Gerth, H. H. and Wright Mills, C. (1948). *From Max Weber: Essays in Sociology*, Chapter 4. Routledge & Kegan Paul, London.

Granovetter, M. (1973). The strength of weak ties. *American Journal of Sociology*, **78**(6):1360–1380.

Heckscher, C. and Donnellon, A. (1994). *The Post-Bureaucratic Organization: New Perspectives on Organizational Change*. Sage, Thousand Oaks, CA, EUA.

Kadushin, C. (1995). Friendship among the french financial elite. *American Sociological Association*, **60**(2):202-221.

Krackhardt, D. (1994). Graph theoretical dimensions of informal organizations. In Carley, K. and Prietula, M., editors, *Computational Organization Theory*, pp. 89–111. Lawrence Erlbaum Associates, Inc.

Mayo, E. (1933). *The Human Problems of An Industrial Civilization*. MacMillan, New York.

Mayo, E. (1945). *The Social Problems of an Industrial Civilization*. Harvard Business School.

Roethlisberger, F. J. (1941). *Management and Morale*. Harvard University Press, Cambridge, MA.

Roethlisberger, F. J. and Dickson, W. J. (1939). *The Early Sociology of Management and Organizations*, volume 5 of Management and the Worker. Routledge.

Simon, H. A. (1962). The architecture of complexity. In *Proceedings of the American Philosophical Society*, volume **106**, pp. 467–482. American Philosophical Society.

Taylor, F. W. (1911). *The Principles of Scientific Management*. Harper & Brothers, New York and London.

Wasserman, S. and Faust, K. (1994). *Social Network Analysis: Methods and Applications*. Cambridge University Press.

Whitehead, T. N. (1938). The industrial worker. In *The ANNALS of the American Academy of Political and Social Science*. Harvard University Press.

CHAPTER **6**

DIFFUSION AND INFLUENCE

The most hateful human misfortune is for a wise man to have no influence.
—*Greek historian Herodotus 484–425 BC.*

Learning Objectives

1. Understand the process of social diffusion.
2. Describe strain theory and its five adaptations.
3. Define an opinion leader, authority figure, and change agent.
4. Identify factors for successful adoption of innovation and ideology.
5. Understand the adoption S-Curve and categories of innovator, early adopter, early majority, late majority, and laggard.
6. Identify Cialdini's six principles of influence and how they interact with social networks.
7. Identify French and Raven's five bases of power and how they interact with social networks.

6.1 APPLICATIONS FOR SOCIAL DIFFUSION

The people in your social network matter. That is, the makeup of an individual's social network (the kinds of individuals in it) also influences the ideas, attitudes, and beliefs of the individual. In Western society, the value we place on individualism causes us to overlook relational factors in social phenomena. This is an essential principle for understanding social diffusion processes in connected groups. Social diffusion is the process by which ideas, innovations, knowledge, disease, and other things spread through a social group. Although we have all heard of peer influence, most fail to consider its impact when exploring policy decisions or group dynamics. Groups are more stable if they have a variety of roles present, including people in

Social Network Analysis: with Applications, First Edition.
Ian A. McCulloh, Helen L. Armstrong, and Anthony N. Johnson.
© 2013 John Wiley & Sons, Inc. Published 2013 by John Wiley & Sons, Inc.

roles designed to encourage socializing, and the presence or absence of particular roles can increase or decrease the likelihood that people engage in threatening behavior such as suicide or civil acts of violence (Palinkas et al., 2004). In this chapter, the social mechanisms for diffusion will be explored and presented in the context of insurgency, industry, health behaviors, and crime.

Many people see *insurgency* as a natural outcome of unresolved social grievances. For example, if a group of people are enslaved, denied civil rights, oppressed, lack basic needs, or are restricted in religious worship they are thought to engage in insurgent activities as a means of fighting back. Following this logic, if government can address legitimate grievances, the insurgency will lack popular support and fail. Unfortunately, this theory does not account for relational aspects of this social process. As a result, many military counter-insurgency efforts in the past decade have failed. The US military did an excellent job of conducting many civil works projects in Iraq to address social grievances following their occupation in 2003; however, an insurgency developed in the face of these efforts that plagued the US military for 7 years. Successful counter-insurgency efforts by the US military were invariably supported with some degree of influence operations to shape group perceptions of activities. From a more complete understanding of social theory, the importance of these activities can be better understood.

In *industry*, many believe that the decision by individuals to adopt an innovation is based on the quality of that innovation. For example, if the device is cheaper, faster, better, or solves a problem, then people will want it. Engineering invariably focuses on improving the innovation. Rarely is the social context considered. Cultural anthropologists have studied the diffusion of innovations for years (Rogers, 2003). However, research has shown that the quality of the innovation is usually not the most important factor in the spread of its adoption. In a similar manner, attitudes toward what are acceptable jobs are influenced by those with stronger connections, but information on where to find a job comes from those to whom we are less closely connected (Granovetter, 1978).

Explanations of why people engage in high risk *health behavior* such as smoking, drug abuse, and unprotected casual sex can also be understood from a social diffusion perspective (Smith and Christakis, 2008). Mercken and colleagues (2010) demonstrated significant affiliation and influence effects on smoking among adolescents. Both of these are relational effects. Valente and colleagues (1997) showed that the underlying social network was important to understanding contraceptive use among women in the Cameroon. Similar studies have demonstrated the importance of social network diffusion in other health applications such as the spread of obesity (Christakis and Fowler, 2007); depression (Rosenquist et al., 2010a); and alcohol consumption (Rosenquist et al., 2010b).

Social diffusion provides an improved understanding of risk factors for criminal behavior. For example, the willingness of young people to engage in crime is highly influenced by their peer group and the lack of male authority figures at home (Glaeser et al., 1996). Traditional approaches that focus on risk at the node level (i.e., age, gender, socioeconomics, drug use, etc.) and neglect relational variables fail to explain why the majority of people characterized by the risk are not affected. In other words, these models have very low explanatory value. Studies of crime in Boston, for example, show that 74% of gun assaults occur in 5% of city block faces and corners (Braga et al., 2010). Furthermore, 75% of gun assaults were committed by less than 1% of the city's youth, who were generally repeat offenders with gang affiliations (Braga et al., 2008). Network models of diffusion have been shown to offer greater explanatory power in the diffusion of crime across social groups (Baerveldt and Snijders, 1994; Houtzager and Baerveldt, 1999).

6.2   STRAIN THEORY

Closely related to network diffusion theory is Robert Merton's Anomie (also known as *Strain*) theory (Merton, 1938, 1949). Merton recognized the impact of social structure on behavior and ideology; however, he did not present a formal social network model of his theory. According to strain theory, social groups will often culturally define goals and acceptable means to achieve those goals. In Merton's illustration of his theory within the context of the "American Dream," he argued that wealth was an American goal and hard work and education were acceptable means to achieve those goals. He further describes the cultural view of people who fail to achieve wealth as lazy, unintelligent, or otherwise defective. Unfortunately, people do not have equal opportunities to achieve goals. Americans born of wealthier parents have greater opportunity to amass wealth than those who are born in poverty. Children of wealthy or highly educated parents have greater access to education. This social inequality across the population places a strain on a segment of individuals. The people will respond with one of five adaptations: conformity, innovation, ritualism, retreat, or rebellion.

The strain adaption is based on their response to the societal goals and means. Individuals who accept both the goals and means will *conform*, continuing their attempts to achieve social goals through acceptable means. These people follow the rules and societal norms. Actors who accept the goals, yet reject the means are labeled *innovators*. These individuals seek alternate means to achieve social goals and typically have a blatant disregard for socially accepted means to achieve those goals. These individuals can significantly enhance creative problem solving and develop novel opportunities for an organization. They are also the individuals most likely to engage in criminal activity. When a poor minority with no education is unable to win a high paying job, the innovator is likely to resort to crime in an effort to achieve success through socially unacceptable means. Likewise, the white-collar worker may resort to fraud or other crimes to gain accelerated social mobility. It is the creativity and simultaneous disregard for the rules that makes this individual successful in criminal activities. Actors who follow the rules, yet reject the goal are adapting to strain with *ritualism*. These people will follow society's rules; however, they have given up hope on attaining the socially established goals. Individuals who have abandoned both the social goal and means make up the remaining two adaptations. People in *retreat* will disengage from social norms, often replacing goals and means with new counter-cultural norms. These are typically homeless, hermits, or severe alcoholics. Actors in *rebellion* will replace goals and means and then actively attempt to change the larger social norms. These individuals are most likely to engage in extremist behavior such as terrorism, racial supremacist groups, subversive political parties, and insurgencies.

People who adapt to social inequality through anything but conformity will experience some degree of cognitive dissonance. Extending the theories on balance that were presented in the previous chapter, this cognitive dissonance results from incompatible views between self-identity and the competing social norm (goal attainment or means). This phenomenon leads to the establishment of new social circles with their own subculture. Innovators who engage in criminal activity or deviant behavior to attain social goals, for example, may develop social ties and eventually social circles with other criminals or deviants (Baerveldt and Snijders, 1994). New social norms governing the means of attaining social goals emerge. Those actors that excel under the new social norms will achieve prestige within the newly defined social circle. Similarly, rebellious actors may define both goals and means within the context of a new social circle. Those actors that epitomize the newly established norms achieve greater self-actualization and status. This is an important mechanism for understanding deviant behavior within the context of social structure.

Strain theory has some limitations as a coherent theory. Due to the diversity of social goals and culturally defined means of attaining those goals, there may exist great overlap between the social circles that both define and attract individuals. This overlap makes it difficult to distinguish salient driving variables in group motivation. "One of the major limitations in previous research, however, is that strains are not adequately and properly measured, failing to assess the effects of duration, magnitude, and subjectiveness of strains on delinquency" (Moon et al., 2008). Understanding the strain theory, in combination with the theory of social link formation presented in Chapter 4, however, are important background for our treatment of diffusion in social networks.

6.3   SOCIAL CONTEXT

Social diffusion is the social context mechanism that governs how people adopt an innovation. In this sense of the word many things can be considered as an innovation. It could be the latest cell phone technology, or it could be a religious ideal, or it could be a moral value/judgment. It could be information or even disease. The term *diffusion of innovation* takes its name from those who first investigated the topic for marketing of new products. In recent years, these same models have been used to study the spread of just about anything through a social network.

The spread of ideology or innovations in a group of people is extremely dependent on the underlying social structure (Rogers, 2003; Carley, 1995, 2001). It is also dependent on the underlying cultural structure (Carley, 1991). For each person, their knowledge, attitudes, and beliefs are a function of both who they know (the social network) and what who they know knows (the knowledge or cultural network). The role of these networks in effecting consensus, belief formation, attitude shifts, knowledge diffusion, collective decision making, cooperation, health, and behavior is well established in the literature (Katz, 1961; Glance and Huberman, 1993; Morris and Kretzschmar, 1995; Rogers, 2003; Carley, 1986, 2001; Friedkin, 2001; Deroïan, 2002; Watts, 2003). From fashion preferences (Aguirre et al., 1988), to willingness to take risks (Pruitt and Teger, 1971), people are influenced by their social network. In general, our ideas, opinions, attitudes, and beliefs are a function of who we interact with, their importance to us, our prior ideas, attitudes, and beliefs, our level of education, new information that we receive, the credibility of that source of information, the emotional content of the message, and the extent to which new information agrees with our existing ideas, attitudes, and beliefs (Ajzen and Fishbein, 1980; Erickson, 1988; Friedkin and Johnsen, 2003).

Our understanding of how the social and cultural structures influence the spread of ideas, attitudes, and beliefs has reached a usable level of maturity. As such, multiple methodologies and models, based on over 60 years of research, exist for tracking, assessing, and using these networks to forecast the spread of ideas, key actors in this spread, and the evolution of beliefs and attitudes.

Some individuals have more influence than others. Within these networks, the attitude and influence of a few individuals can radically alter the collective opinions and actions of the group. That is, some individuals, by virtue of their position in the social network, have disproportional influence (Friedkin and Johnsen, 2003; Coleman et al., 1957). Such individuals are often referred to as *opinion leaders* or super-empowered individuals or key actors. Lazarsfeld and Katz (1955) suggested a two-step flow model of the impact of media on social behavior in which mass media information is channeled to the "masses" through opinion leaders. In areas where access to mass media is reduced, due to literacy or the cost of accessing the media (lack of Internet penetration, etc.), those with better access to the media and who have a more literate understanding of media content are likely to explain and diffuse the content to others and so have a disproportional influence on changes in ideas, attitudes, and beliefs.

A variation on the opinion leader is the role of the authority figure. In the famous Milgram experiment (Milgram, 1963, 1974) conducted at the Yale University, Milgram studied the willingness of people to obey perceived authority figures, when their instructions conflicted with moral conscience. He found that 26 of the 40 subjects (65%) followed the immoral instructions. This suggests that people can be influenced to a greater degree, by a minority of people, if the minority is seen as an *authority figure*. In some sense the authority figure is very similar to a node with high prestige within the context of a particular social circle. A key point is that the opinion leader and authority figure may not be obvious. An important step in understanding and assessing organizational risk is to realize that opinion leaders, authority figures, and high prestige nodes exist and have high influence on group behavior.

Chapter 4 presents several theories pertaining to social link formation and the development of social norms and mores. These group values will of course vary from group to group. Individuals who epitomize these group values will hold positions of high status or prestige in the group. These individuals will be able to influence others in the group by virtue of their status. In the context of diffusion, these individuals are known as *opinion leaders*. Opinion leaders will occupy positions of high centrality in a social network. In general they will have high degree and high closeness centrality and will be able to directly influence many in the group.

An opinion leader must be very careful of what innovations or ideology they adopt. If an opinion leader adopts something that contradicts a group value, they risk losing their status as an opinion leader by no longer epitomizing social norms. Therefore, successful change must be implemented slowly and with the group consensus.

Several factors contribute to successful adoption by group members (Rogers, 2003). First the innovation must have a *perceived benefit* over the existing status quo. Note that this is not an actual benefit, but there must exist some incentive for some one to change the existing status quo. The innovation must be *compatible* with existing beliefs and systems. People may refrain from buying the latest new video game if their computer system cannot support the memory and processor requirements. In this instance, the innovation is not compatible with the existing system. There must exist an extreme perceived benefit to motivate the consumer to buy an upgraded system. Likewise, if a radical idea contradicts a moral value held by the group, a compelling case (perceived benefit) must be made to justify the idea. The *complexity* of the innovation must be low. If the innovation requires too much effort to adopt, it will not be as successful. *Trialability* is another important consideration. If people can test out a new innovation prior to making a decision to adopt permanently, they will feel more comfortable in their decision.

All of these considerations still speak to the quality of the innovation. The social structure is perhaps an even more significant factor however. Recall that the very source of an opinion leader's influence is in their conformance to group norms and mores. *Change agents*, by contrast, are individuals who attempt to change the ideology or values of a group or introduce new technology. By definition, they are not similar to the group by nature of their desire to implement change. Successful change occurs when a change agent is able to influence opinion leaders to adopt the innovation, at which point it can spread to the rest of the social group. Reviewing the causes for social links, change agents must make a deliberate effort to establish homophily, reciprocity, proximity, and balance with opinion leaders in order to influence them to change. The change must be measured and incremental, so that their intended changes are not completely unappealing to the opinion leader. Finally, the change agent must consider perceived benefits, compatibility, complexity, and trialability when implementing change.

Check on Learning

What are the factors that influence the spread of a new innovation and its adoption?

Answer

Spread of a new innovation is influenced by the underlying social and cultural structures. This includes who and what they know, and also what is known by who they know. The adoption of new innovations is affected by its perceived benefit, compatibility with existing beliefs and structures, minimum complexity, and trialability by testing.

In Example 1.2 on page 13, a military noncommissioned officer assumed leadership of a military unit. Although he had formal authority over the unit, his effectiveness was compromised by his failure to recognize a low ranking, yet informal leader within the unit. In his previous assignment, which was very similar to the present one, he had been able to implement many successful changes leading to his promotion and current assignment. In the present unit, none of these measures was effective, due to the animosity between him and the informal leader. This example shows that even military discipline cannot compensate for the strong informal network dynamics. Even formal managers must assess the organizational culture and informal opinion leaders within their organizations. Successful change is implemented in cooperation with the opinion leader.

Example 6.1

Chem Coy

Chem Coy is the anonymous name of a large chemical manufacturing company in Western Australia (Alexander et al., 2011). The company attempted to implement a new enterprise resource planning (ERP) tool to improve efficiency and reduce costs. The ERP was a well-validated tool and a proper feasibility study was conducted to determine that the ERP would improve profits if implemented. Consultants were brought in to manage the organizational change as Chem Coy adopted the new system.

The executive level leadership all discussed the new tool and had opportunities to engage in discourse, ask questions, and plan for the ERP implementation. Through these discussions, the executive level leadership reached consensus on the efficacy of the ERP and expressed positive approval of the new tool. Some middle-management and lower-level employees were not consulted on the ERP implementation and were simply told that they would receive training and be required to start using the new system according to a schedule created by the consultants.

The implementation of the new ERP system was considered a failure. One year following the failed implementation, the company was no longer deemed to be competitive, the executive level leadership was laid off, and Chem Coy was merged with its parent company. Subsequent relational algebra (Chapter 7) revealed an informal network in the company's production chain. Central figures in the informal network were not only ignored throughout the implementation of the ERP, they had outspoken objections to the new system that were never addressed. This provides another example where formal authority may not overcome the lack of ideology diffusion throughout the informal network. Perhaps, the ERP implementation would have been effective had management identified informal opinion leaders, sought their buy-in by investing time and resources listening to their concerns and explaining the benefits of the new ERP. These examples demonstrate the importance of social context in diffusion.

Check on Learning

True or False. Individuals who attempt to change the ideology or values of a group or introduce new technology are known as *opinion leaders*.

Answer

False. These are known as *change agents*.

6.4   GROUP IMPACTS ON DIFFUSION

Tightly connected groups are prone to group think. In addition to the impact of the opinion leader there is a social network group effect. That is, research shows repeatedly that groups can exert implicit pressure to influence opinions leading to higher conformity, more extreme views, and a group-think mentality where contradictory evidence is ignored. As discussed in Chapter 4, Asch's study demonstrates how people will ignore first-hand evidence and conform to the common opinion held by the group (Asch, 1955, 1956). Furthermore, Asch found that a single voice among the group that was not unanimous resulted in a drop in conformity from 37% to 5.5%. Asch's finding provided empirical evidence that an individual's perception of reality may be influenced by those with whom they share social connections. This conformity effect increases as people have increased dependence upon the group for social acceptance and when the correct solutions to problems become less clear.

Cliques, isolated tribes, and groups that inhibit interaction with the outside are all more prone to this negative group-think side of networks. This concept is sometimes referred to as *social insulation*, taking its name from a physics analogy. Thermal insulation works by creating an evacuated space between molecules of high and low temperature, thereby preventing heat transfer. By analogy, a radical idea or innovation is like heat energy that is diffused through a substance through interacting molecules. When there is an evacuated space preventing molecules from interacting, the heat cannot diffuse and the substance will remain hot. With a social network, a radical idea presented to an isolated group has limited resources to verify the veracity of the radical idea. If they see that others may have similar views, Asch's conformity suggests that they are likely to conform to the group's opinion. In contrast, an individual with access to others outside the group with differing opinions has greater freedom in questioning the radical idea. They also have a social freedom to associate with a different social circle that is more compatible with their views. A cult, such as the Branch Davidians in Waco, was able to maintain radical beliefs, because their social network was isolated from the rest of society, and highly connected with those of similar beliefs. Terrorist groups are able to influence insurgents through a social network designed to isolate the insurgent in a group of radical Islamists (Howard and Gunaratna, 2006). In these examples, social insulation is necessary to heat the group views with a radical ideology.

People are more inclined to adopt group beliefs than reality. Individuals need social acceptance as explained in Chapter 4. This inherently requires some level of social conformity. When an actor perceives that group views and his observation are inconsistent, they experience cognitive dissonance resulting from the unbalanced triad. Nyhan and Reifler (2010) demonstrate this concept in politics where they show that citizens will reject certain facts when they contradict political group values. A 2011 poll of US voters showed that three out of four Republican voters thought that the US President could take greater action to reduce gasoline prices, in contrast to one out of four Democrat voters with the same view. In 2006, when there was a Republican president instead of a Democrat, however, the views were reversed. Only one out of four Republican voters thought that the US President could do more to affect gasoline prices, whereas three out of four Democrat voters held the same view. Arguably, the US President has limited ability to affect global gasoline prices and US policies are not significantly different between 2006 and 2011. Why then, would voter views be so different? Nyhan and Reifler's explanation is that when a person holds an affinity for a particular politician and perhaps gains some identity through the represented political party, it is difficult to maintain a view that they are not performing well. This creates a cognitive dissonance. It is easier for the individual to reject the facts in order to support an opinion that provides them identity and acceptance. Thus, the greater extent to which a fact may contradict social norms and values, the more likely an actor will reject the fact in favor of the group belief.

The actual benefit or value of the innovation is not as important as the social structure through which it is diffused. In the 1940s, W. Edward Deming proposed total quality management (TQM), which was largely ignored in the United States. Deming lacked any prestige or opinion leadership at the time. Deming eventually met a leading Japanese businessman, Ichiro Ichikawa, who had a high level of opinion leadership within Japan, whom he convinced of the value of TQM. Through Ichikawa's social network, TQM rapidly diffused and eventually revolutionized quality engineering in Japan, as well as the United States, where it had previously been ignored (Halberstam, 1986).

Another example of failed innovation was evidence that lime juice cured scurvy. James Lancaster, a British sea captain, proposed this innovation in 1601 with little effect. The innovation was rediscovered almost a century and a half later in 1747 by an English Navy physician, James Lind. Neither Lancaster nor Lind held any opinion leadership within their social networks and the innovation was not adopted. It was not until 1795 that scurvy was eradicated throughout the British Navy with the innovation of lime juice (Mosteller, 1981).

Wellin (1955) conducted a 3-year ethnographic study of a failed public health effort to convince families in Los Molinos, Peru, to boil their drinking water to prevent water-borne illnesses that were plaguing the town. A young health professional, Nelida, was sent to educate housewives on the benefits of boiling water. She succeeded in convincing 11 out of approximately 200 families to boil water. Wellin reports that boiling water had a cultural meaning within the target society. Certain objects were considered hot or cold. Water was cold. "Cooking" the water had been culturally linked to illness and members of that society had been taught to detest boiled water. In fact, one woman who adopted the innovation of boiling water became ostracized by other housewives of Los Molinos for her decision to contradict the social norm. In this situation, the reality of water-borne illness was rejected in favor of the cultural belief.

6.5   NETWORK STRUCTURE AND DIFFUSION

Rogers mapped the diffusion of innovations as the rate of adoption. He observed that the adoption process followed an S-shaped curve, where at a given point in time, the number of adopters rapidly increased and then tapered off after most available people adopted the innovation. He discovered that the rate of adoption over time (the first moment of the adoption curve) approximated the bell curve of the normal probability density function. Rogers categorized the bell curve based on standard deviations from the mean adoption time. The five categories consist of innovators (<2 standard deviations before the mean); early adopters (between 1 and 2 standard deviations before the mean); early majority (between the mean and 1 standard deviation before the mean); late majority (between the mean and 1 standard deviation after the mean); and laggards (later than 1 standard deviation after the mean).

Moore (1991) describes a "chasm" between the early adopters and the early majority to describe the point at which many innovations fail to diffuse. Rogers et al. (2005) further defined a critical threshold, also known as a "tipping point" (Gladwell, 2002) where a sufficient number of adopters are required to generate mass adoption or the death of the innovation. If enough actors adopt the innovation, it may diffuse virally throughout the social network, reaching a saturation point where new adopters are unlikely. Eventually, adoption will fade as new technologies and ideas replace the existing innovation. When introducing an innovation or idea, there are opportunities to affect attitudes toward the innovation up to the tipping point. Past this point, when the innovation is spreading viral, there is little ability to adjust people's perception of the ideology or innovation. There may be opportunity to introduce alternatives that may lower the saturation point or create conditions to induce an earlier drop-off point where the adoption fades. Figure 6.1 shows Rogers' diffusion curve with Moore's chasm.

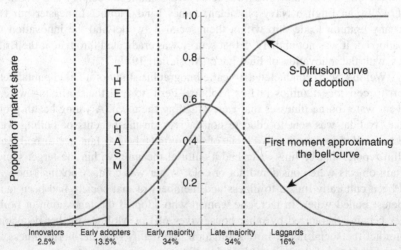

FIGURE 6.1   Moore's chasm and Rogers' diffusion curve [after Moore, 1991; Rogers et al., 2005].

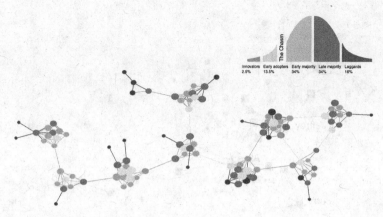

FIGURE 6.2   Actors involved in the diffusion process.

Social network analysis provides tools to map the social structure for which innovations will diffuse. Innovators will occupy boundary spanning, high brokerage positions between network subgroups. Innovators are often not central to a social circle. Centrality is typically correlated with conformity to social norms. Either central figures hold prestige, in part based on their epitomizing social norms or other actors in the network conform to the opinion leader due to their influence. This mechanism creates group think, which actually inhibits innovation. Nodes that span multiple social circles are in a position to diversify their need for acceptance across multiple groups, which frees them to take risk. They are also in a position to broker successful innovations from one social circle to another. In contrast, actors who affiliate with a limited set of social circles are less likely to become aware of potential innovations. Figure 6.2 displays network positions of actors involved in the diffusion process.

Boundary spanning nodes are not in a position to diffuse an innovation by themselves. The innovator must have a social connection with an opinion leader within a social circle in order to diffuse the idea or innovation. Opinion leaders are often central figures within the network. High degree and closeness centrality are good indicators of opinion leadership. Opinion leaders in Figure 6.3 are represented with a star. Those actors closest to the opinion leader make up the early majority. Innovators are squares. Laggards occupy positions on the edge of a social network with no position of brokerage.

An interesting extension of diffusion is a competing influence model. Consider two actors in the network displayed in Figure 6.3: Robert and Fred. Each actor has an opposing ideological belief. For example, Robert believes in government regulation of the economy and Fred believes in free markets. Jeffrey, Alice, and Jerry are in a position to establish the social policy for this notional government. Within this network, Robert is more likely to diffuse his ideology to Jeffrey through the shared connection they have to an innovator. Similarly, Fred is more likely to influence Alice. Jeffrey and Alice share connections with five common alters. Depending upon the position of those five alters on the issue Jeffrey and Alice may be more or less likely to adopt a position. Valente (1996) argues that an

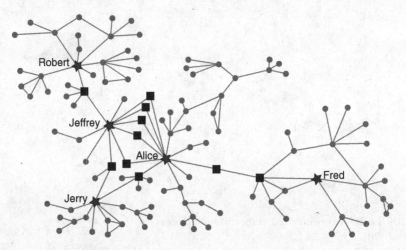

FIGURE 6.3   Opinion leaders in a network.

actor's likelihood to adopt an innovation is dependent upon the number of alters that have also adopted the innovation. Thus one might argue that if Robert is faster in diffusing his ideology to Jeffrey, who in turn spreads the ideology to the alters he shares with Alice, Alice may be effectively inoculated to Fred's ideas. Fred must now overcome the social norm emerging within Alice's social circle. Assuming that Robert and Fred are effective at convincing Jeffrey and Alice to adopt their respective views, they must in turn diffuse that ideology to Jerry. Jerry shares a single common alter with Jeffrey and similar connection to Alice. In this situation, neither Jeffrey nor Alice is able to gain a consensus advantage for influencing Jerry. Jerry is then socially free to weigh the merits of both ideologies against his current socially accepted values to determine which one he may adopt. Individuals will apply non-network based criteria in making an adoption decision as well. Actors evaluate their perceived value of the innovation, the compatibility with existing ideology or technology, the complexity of the innovation, the perceived risk and commitment. The social network is simply another competing factor in a person's adoption decision. Social pressure is an important and often overlooked factor in social diffusion. The social network may also vet the trust an actor places on the information they receive from others. Alice is more likely to believe the information coming from the shared connection she has with Fred than a non-vetted source. If that information exists outside of Alice's social norms, however, it is not likely that she will adopt the ideology as it would create a cognitive dissonance with group values. Thus, even nonnetwork based criteria can be influenced by social network position.

6.6   GROUP INFLUENCE STRATEGIES AND BASES OF POWER

The discussion thus far has been centered on the dynamics of how influence behaves in networks. This section will identify well-established strategies for group influence, often used in marketing as proposed by Cialdini (2001), or in leadership as proposed by French and Raven (1959). These strategies will be tied back to the network based mechanisms presented earlier in the chapter. Cialdini identifies six principles of influence: reciprocity, commitment, social proof, authority, liking, and scarcity.

Reciprocity, as we have seen in Chapter 4, is a means to establish a relationship. People not only tend to reciprocate friendship, they also reciprocate a favor. This has led to the use of free samples in marketing. In conferences, Cialdini often uses the example of Ethiopia providing humanitarian aid to Mexico after their 1985 earthquake, even though Ethiopia was suffering from a devastating famine and civil war. Ethiopia was reciprocating diplomatic support provided by Mexico after Italy invaded in 1935.

Individuals who commit to an idea, contract, or goal are more likely to honor it, even if the original incentive has been removed. An example is a salesman who either raises the price of a product or discloses hidden fees at the last minute. This technique works, because the individual has already committed to buying the product. The commitment is actually a process where the person has associated the product with their self-identity. They believe that the product may provide greater group acceptance or there may exist a cultural norm to honor commitment. Cialdini ties his definition of commitment to being congruent with self-image. In Chapter 4, self-image was in turn placed in the context of a wider social circle and acceptance. Thus, commitment and self-image may be impacted by network effects.

Social proof is a concept where people do things that others do. Cialdini aligns this back to Asch conformity. This principle is directly related to social acceptance and link formation. People have a tendency to conform to perceived social norms.

Authority has also been presented in this text, earlier in this chapter. Cialdini cites incidents such as the Milgram experiment or the My Lai massacre in the US Vietnam War. People have been demonstrated to conduct even unethical behavior, when directed by a perceived authority figure. The authors posit that the authority figure is actually a group opinion leader. The opinion leader may derive their opinion leadership from a perceived status, expertise, or social prominence within an individual's group. However, it is the individual's desire to gain and maintain membership and acceptance within a social circle that causes them to follow opinion leaders. This modification of Cialdini extends this principle to peer pressure and group think.

People are influenced by people they like. Cialdini shows that people are more likely to buy something if they like the person who is selling it to them. An extension of this is that people who are physically attractive are more likely to be well-liked (Miller, 1970; Dion et al., 1972). People prescribe positive cultural attributes to attractive individuals without realizing it. There are other reasons for individuals to be liked by others, such as their epitomizing cultural norms, which in turn leads to their opinion leadership. Liking is more distinct from opinion leadership in situations where there is an initial or temporary contact with another, such as buying a car. It is unlikely that the customer will ever see the salesperson again after the purchase. However, in this situation, there may still be a desire for social acceptance on the part of the customer, which may make the customer more susceptible to influence by the salesman.

Scarcity is Cialdini's final principle. He explains that offers available for a "limited time only" will make a person more likely to buy. The perception of scarcity will instill a sense of urgency on the part of the buyer, making them more likely to purchase in fear of "missing out." The authors argue that there must be some other perceived benefit of the product to generate sales. The benefit, if it provides value in a social context, will be a source of acceptance. An individual may normally wait or negotiate to obtain the product at the lowest possible cost, optimizing their benefit. However, if the product is scarce, their desire to have the product outweighs the risk of overpaying for it. Their fear of missing the opportunity outweighs their desire to negotiate for a better price, or taking time to see if others in the social group will adopt the innovation.

French and Raven (1959) identify five bases of power used by leaders to influence others: coercion, reward, legitimate, expert, and referent. *Coercion power* is the ability of a person to force someone to do something that they do not want to do. This often involves threats or abuse. In a military context, it is the application of combat force upon another population. A leader employing this type of power may threaten to reduce an employee's hours or withhold a bonus. The only means of influence that coercion can achieve is that of short-term compliance. Over time, if coercion is the primary means of influence, if it is applied unfairly across the group, or if it is perceived to be abusive, people will be motivated to resist and undermine that leader. This type of power completely ignores the informal network. It may expect people to conduct actions outside the cultural norm, which may threaten their acceptance and sense of identity. It is not surprising that this type of power is often ineffective in the long run for achieving influence goals.

Reward power is the ability to provide a person something they desire or take away a negative stimulus that they dislike. Rewards are generally more effective than coercion in motivating compliance. Unfortunately, rewards typically need to be greater each time they are applied to have the same influence effect. When rewards no longer have a perceived value, people are no longer influenced by the reward. The perception of value is only understood within the context of the social circle for which it is applied. Thus, if everybody receives a bonus, it becomes a welcome expectation rather than a reward to influence behavior. Opinion leaders are important informal leaders who set value to potential rewards. If the group believes that rewards are being applied in a manner that is unfair or given to people not esteemed by the informal group, reward power can have a negative effect and lead toward group resentment and resistance.

Legitimate power is the ability to instill a sense of obligation or responsibility based on the formal position an individual occupies. This power only lasts as long as individuals occupy the same formal positions. People often expect formal leaders to exercise some degree of coercion and reward in the execution of their duties (Bass, 1990), which creates an interaction effect between legitimate, reward, and coercion power. If the informal group perceives the legitimate power to be abused or applied unfairly, however, it can become an ineffective means of influencing others. Sociologically, the basis for legitimate power is drawn from two complementary theories. The legitimate power is always held by a perceived authority figure and is supported by the Milgram experiments. Secondly, a formal leader's role is legitimized by a culturally accepted view of the social circle that provides the formal leader with his position. Thus, rejecting the formal leader may contradict cultural norms and mores and threaten an individual's acceptance in the social circle. Leaders should recognize that their formal position provides influence only in as much as organizational culture places value on their formal position. Opinion leaders not only have the ability to reinforce the formal leader's position; they have equal ability to undermine the leader and strip him of his legitimate power, even if he retains the formal position.

Expert power is the ability to provide others with access to knowledge, resources, specialized tasks, information, or other forms of expertise. For example, a novice going on a white water rafting trip will listen to (be influenced by) the guide due to his expertise associated with white water rafting. Similarly, an employee may defer to the judgment and guidance of a more experienced colleague. This power, combined with the correct use of reward power develops trust by those who are influenced. The expert power is again based on perceptions held by those in a culturally defined social circle. Some individuals may initially have high expert power based on observed traits such as an advanced degree, military rank, or specialized training. If, however, those individuals fail to demonstrate expertise, make consistent mistakes, or behave in a manner that contradicts the culturally defined norms of their expertise (i.e., a military officer who is physically unfit), they can lose their expert power in the view of those they are trying to influence.

Referent power is the ability to appeal to an individual's sense of personal acceptance or approval. In other words, referent power appeals directly to a person's sense of self-worth. This type of power is derived from position in the informal network. Formal leaders may attain this type of power by engaging in the informal network or demonstrating that they adhere to the culturally defined norms and mores of the group. This power is often manifest as charisma or charm (Raven, 1990). The referent power leader assumes a position as a role model. This type of power is highly effective when combined with other forms of power.

In all cases, however, the ability of an individual to apply any of these bases of power is dependent upon the informal social network and culturally defined norms. These forms of power can be linked to their effectiveness in attitude change. Kelman (1958) identified three types of attitude change: compliance, identification, and internalization. Compliance occurs when a person's behavior is not necessarily linked to their beliefs. An individual complies to avoid punishment, gain reward, or achieve social acceptance. Identification involves a change in belief in an effort to become more similar with someone the person likes or admires. The desire to adopt a new idea is not based on the content of the idea, but on the desire to have a relationship. Identification is an effect of cognitive dissonance in maintaining a positive relationship with an alter who possesses certain beliefs. The beliefs rarely persist beyond the relationship with the alter. Internalization occurs when a belief is intrinsically accepted and merged with the existing values. The content of the belief is considered desirable to the individual and persists beyond relationships with alters.

Coercion power only has the ability to achieve compliance in the short term. Over time, the target population will begin to resist if this is the only base of power applied. Reward power is more effective at achieving compliance than coercion power; however, overtime rewards must be greater and greater. This makes reward power only effective in the short term as well. Legitimate power is necessary to move a person toward identification. However, the legitimate power must be approved of by the cultural norms of the individual's social circle. Leaders who abuse their legitimate power can quickly lose their influence and the target audience will readily move toward compliance as they lose respect for the leader. Expert power is the first base of power that may solicit emulation and begin to broach internalization. However, mistakes or actions that violate expected norms of the expert may cause them to lose the ability to maintain internalized values in a target audience. Thus, expert power only reaches the cusp of the identification–internalization threshold. Referent power is necessary to reach internalization. Referent power appeals directly to a person's core sense of self and as such has greater ability to gain trust and influence an individual's value system. Once an agent internalizes a belief, they are much less likely to change even if the relationship no longer persists. Recall that referent power is gained through direct engagement with the informal social network. It requires behavior consistent with culturally defined norms and relationships with key opinion leaders in the social circle.

Check on Learning

Identify the types of power reflected in the following situation:

You are a CEO of a large corporation and your medical practitioner tells you that your blood pressure is too high and if you do not change your lifestyle you will have a heart attack. She advises you to take medication to reduce your hypertension and urges you to replace your current high pressure job with one that is less stressful, enabling you to play more golf, which you really like. You obtain much social acceptance and respect from your position and do not wish to leave this prestigious job.

Answer

1. Expert power as a medical doctor has an authority and knowledge not usually held by others.

2. Coercion power as the doctor is threatening you in order to convince you to do something you do not want to do.

3. Reward power may also be a possible power type reflected in this situation as the reward is playing more golf and not having a heart attack—illustrated by the removal of the individual from the location that is causing the problem.

6.7   SUMMARY

Networks are ubiquitous. As we have seen, social and cultural networks influence the spread of ideas, beliefs, and attitudes in numerous ways: opinion leaders, hidden sources of authority, group think, composition, and topology. Understanding how social and cultural networks affect the spread of ideas, beliefs, and attitudes is critical to developing an effective strategy to influence a social group. This understanding provides insight into new policy interventions that may prove effective in marketing, managing change, winning asymmetric conflicts, and reducing crime. An ability to understand how ideology spreads through a social network can provide critical insight to industry leaders, military commanders, law enforcement professionals, and organizational planners. This understanding will allow planners to develop improved strategies for influencing social networks through a variety of means.

In applying network principles to counter-insurgency, it is important to balance an understanding of social diffusion with actual and perceived grievances. The first task should be to identify opinion leaders in the social group. The next step is to identify the perceived social grievances according to the opinion leaders. At this point, a change agent, possibly a military leader, can assess what grievances they can address and how they can address these issues. This should involve discussion between the change agent and the opinion leader. As the change agent develops their intervention, they must keep in mind the four factors of successful adoption: perceived benefit, compatibility, complexity, and trialability. It is important to have the opinion leader's buy in to implement any type of social change. Although the change agent needs a reliable product to sell the social group, they must understand that the opinion leader's buy-in is the most critical component to actually implementing and change.

A manager in industry must also recognize the influence of the informal network and his ability to lead depends upon his successful interaction or participation within that network. If we consider a definition of leadership as "convince others to accept your goals as their own," then a manager's leadership is based not on orders, coercion, and reward, but rather on influence. What source of influence does a manager have? They may have authority or expertise, which may affect their perceived status within the group. There may also be some status ascribed to the rank or status of the position if it is a culturally defined goal of the workplace social circle. If he has some opinion leadership, he may be able to alter the culturally defined goals to make them group performance standards. Involving members of the group in defining goals is an excellent technique to develop consensus; however, a manager with low opinion leadership can easily lose control of this process. Alternatively, the manager can identify the informal opinion leaders through social network analysis; develop relationships with those individuals, understanding the development of social links from Chapter 4; and then convince the opinion leaders to promote the manager's views. Thus, the successful manager must recognize the informal network and then decide to join this network or engage the opinion leaders to influence the group.

An important application area of influence is in marketing. The principles outlined in this chapter demonstrate that the informal social network is more important for marketing an innovation than the actual quality of the innovation itself. Pfeffer (2008) presented an excellent example of marketing a German novel, Das Kind, at the Sunbelt conference. The novel was a thriller, so Pfeffer mapped out a network of bloggers throughout Germany. He modeled several diffusion strategies, concluding that it would be more effective to concentrate resources in one cluster of the network. He designed a mixed-reality online game that served as a prequel to the book. They sent a pizza with a thumb drive taped to the box to 12 central individuals of whom 5 actually put the thumb drive in their computer. The game quickly diffused throughout Germany in the span of a couple months. Two weeks prior to the novel's release it was the twenty-third best seller on Amazon through presales, demonstrating that the calculated diffusion through the network was perhaps more important than the quality of the novel, as no one had read it yet.

Cialdini's six principles of influence for marketing are also dependent upon the informal social network. Crime can be considered an innovation that is diffused. As proposed by strain theory, social inequality exists where there is disparate access to wealth and education. Innovators who reject the acceptable means of attaining wealth and explore crime as an alternative still require social acceptance. These innovators can diffuse crime through opinion leaders or establish new social circles where crime is socially acceptable. Those who excel in criminal activities may gain prestige and become opinion leaders in newly defined cultures where certain crime is acceptable. Strategies to counter crime have focused on removing the strain by providing opportunity to the disadvantaged. This is similar in practice to addressing grievances in counter-insurgency. Strategies must address the social network. Opportunities must be diffused through culturally defined opinion leaders. These opportunities must also be presented in a manner that is culturally acceptable given the norms and mores of the unique social context of the disadvantaged population.

Social networks provide an important context for diffusion and influence. Any effort to influence people must consider the informal network and its unique culture. From insurgency to healthcare, leadership to crime, marketing to politics, the social dynamics of networks are present. This chapter has provided a brief summary of important theories in diffusion and influence and integrated them through a context of social networks.

EXERCISES

6.1. List five recent electronic devices that have hit the market. Which of these devices have you adopted? How soon after they were released did you adopt them? Did you buy them because someone you respect had purchased them? Write down where you fit in Rogers' adoption categories (innovators, early adopters, early majority, late majority, laggards).

6.2. Compare and contrast the roles of the opinion leader, authority figure, and change agent.

6.3. Find examples from your own social networks of the application of Cialdini's six principles of influence.

6.4. You are the manager of two project teams whose performance needs to be raised. You wish them to adopt a new IT system, which will enable them to be more productive as well as result in greater customer satisfaction. Using the models on diffusion and adoption and types of power discussed in this chapter explain how you would achieve your goal.

6.5. Who, in your social networks, is an opinion leader? How much do you trust new information or ideas they give you? How readily do you accept these new ideas when they do not conform to your existing ideas and attitudes?

REFERENCES

Aguirre, B. E., Quarantelli, E. L., and Mendoza, J. L. (1988). The collective behavior of fads: the characteristics, effects, and career of streaking. *American Sociological Review*, 53(4):569–584.

Ajzen, I. and Fishbein, M. (1980). *Understanding Attitudes and Predicting Social Behavior*, volume 278, Englewood Cliffs, NJ: Prentice-Hall.

Alexander, P., Armstrong, H., and McCulloh, I. (2011). Towards supply chain excellence using network analysis. *IEEE Network Science Workshop*.

Asch, S. E. (1955). Opinions and social pressure. *Nature*, 176:1009–1011.

Asch, S. E. (1956). Studies of independence and conformity: I. a minority of one against a unanimous majority. *Psychological Monographs: General and Applied*, 70(9):1–70.

Baerveldt, C. and Snijders, T. (1994). Influences on and from the segmentation of networks: hypotheses and tests. *Social Networks*, 16(3):213–232.

Bass, B. (1990). *Bass & Stogdill's Handbook of Leadership (3rd ed.)*. Free Press, New York.

Braga, A.A., Hureau, D., and Winship, C. (2008). Losing Faith? Police, Black Churches, and the Resurgence of Youth Violence in Boston. *Ohio State Journal of Criminal Law*, 6:141–172.

Braga, A., Papachristos, A., and Hureau, D. (2010). The concentration and stability of gun violence at micro places in boston, 1980–2008. *Journal of Quantitative Criminology*, 26:33–53. doi: 10.1007/s10940-009-9082-x.

Carley, K. (1991). A theory of group stability. *American Sociological Review*, 56(3):331–354.

Carley, K. M. (1986). Knowledge acquisition as a social phenomenon. *Instructional Science*, 14:381–438. doi: 10.1007/BF00051829.

Carley, K. M. (1995). Communication technologies and their effect on cultural homogeneity, consensus, and the diffusion of new ideas. *Sociological Perspectives*, 38(4):547–571.

Carley, K. M. (2001). Learning and using new ideas; a sociocognitive perspective. In Council, N. R., editor, *Diffusion Processes and Fertility Transition: Selected Perspectives*, chapter 6, pp. 179–207. National Academy Press, Washington DC.

Christakis, N. and Fowler, J. (2007). The spread of obesity in a large social network over 32 years. *New England Journal of Medicine*, 357(4):370–379.

Cialdini, R. B. (2001). *Influence: The Psychology of Persuasion*. Collins Business Essentials, New York, NY.

Coleman, J., Katz, E., and Menzel, H. (1957). The diffusion of an innovation among physicians. *Sociometry*, 20(4):253–270.

Deroïan, F. (2002). Formation of social networks and diffusion of innovations. *Research Policy*, 31(5):835–846.

Dion, K., Berscheid, E., and Walster, E. (1972). What is beautiful is good. *Journal of Personality and Social Psychology*, 24(3):285–290.

Erickson, B. H. (1988). The relational basis of attitudes. In Wellman, B. and Berkowitz, S. D., editors, *Social Structures: A Network Approach*, *volume 2 of Structural Analysis in the Social Sciences*, pp. 99–121. Cambridge University Press, New York, NY.

French, J. and Raven, B. (1959). The bases of social power. In Cartwright, D. (Ed.), *Studies in Social Power*. Institute for Social Research, Ann Arbor, MI.

Friedkin, N. E. (2001). Norm formation in social influence networks. *Social Networks*, **23**(3):167–189.

Friedkin, N. E. and Johnsen, E. C. (2003). Attitude Change, Affect Control, and Expectation States in the Formation of Influence Networks. *Advances in Group Processes*, **20**:1–29.

Gladwell, M. (2002). *The Tipping Point: How Little Things can make a Big Difference*. Little, Brown and Co, Boston, MA.

Glaeser, E. L., Sacerdote, B., and Scheinkman, J. A. (1996). Crime and social interactions. *Quarterly Journal of Economics*, **111**(2):507–548.

Glance, N. S. and Huberman, B. A. (1993). The outbreak of cooperation. *The Journal of Mathematical Sociology*, **17**(4):281–302.

Granovetter, M. (1978). Threshold models of collective behavior. *American Journal of Sociology*, **83**(6):1420–1443.

Halberstam, D. (1986). *The Reckoning*. New York, NY: Avon Books.

Houtzager, B. and Baerveldt, C. (1999). Just like normal: a social network study of the relation between petty crime and the intimacy of adolescent friendships. *Social Behavior and Personality an International Journal*, **27**(2):177–192.

Howard, R. and Gunaratna, R. (2006). Winning the war on terrorism in singapore. *Lecture given at the United States Military Academy Combating Terrorism Center*.

Katz, E. (1961). The social itinerary of technical change: two studies on the diffusion of innovation. *Human Organization*, **20**(2):70–82.

Katz, E. and Lazarsfeld, P. F. (1955). *Personal Influence: The Part Played by People in the Flow of Mass Communication*. The Free Press, Glencoe, Ill.

Kelman, H. C. (1958). Compliance, identification, and internalization: three processes of attitude change. *The Journal of Conflict Resolution*, **2**(1):51–60.

Mercken, L., Snijders, T., Steglich, C., Vartiainen, E., and de Vries, H. (2010). Dynamics of adolescent friendship networks and smoking behavior. *Social Networks*, **32**(1):72–81.

Merton, R. (1938). Social structure and anomie. *American Sociological Review*, **3**(5):672–682.

Merton, R. (1949). Social structure and anomie: Revisions and extensions. In Anshen, R., editor, *The Family*, pp. 226–257. Harper Brothers, New York.

Milgram, S. (1963). Behavioral study of obedience. *Journal of Abnormal Psychology*, **67**:371–378.

Milgram, S. (1974). *Obedience to Authority: An Experimental View*. New York, NY: Harpercollins.

Miller, A. G. (1970). Role of physical attractiveness in impression formation. *Psychonomic Science*, **19**(4):241–243.

Moon, B., Blurton, D., and McCluskey, J. D. (2008). General strain theory and delinquency: focusing on the influences of key strain characteristics on delinquency. *Crime & Delinquency*, **54**(4):582–613.

Moore, G. A. (1991). *Crossing the Chasm*. Harper Business, New York, NY.

Morris, M. and Kretzschmar, M. (1995). Concurrent partnerships and transmission dynamics in networks. *Social Networks*, **17**:299–318. Social networks and infectious disease: HIV/AIDS.

Mosteller, F. (1981). Innovation and evaluation. *Science*, **211**(4485):881–886.

Nyhan, B. and Reifler, J. (2010). *Misinformation and Fact-Checking: Research Findings From Social Science*. Washington, DC: New America Foundation.

Palinkas, L. A., Johnson, J. C., Boster, J. S., Rakusa-Suszczewski, S., Klopov, V. P., Fu, X. Q., and Sachdeva, U. (2004). Cross-cultural differences in psychosocial adaption to isolated and confined environments. *Aviation, Space, and Environmental Medicine*, **75**(11):973–980.

Pfeffer, J. (2008). The structure of buzz: Modeling rumor diffusion in dynamic social networks. *Sunbelt XXVIII International Social Network Conference. International Network for Social Network Analysis*.

Pruitt, D. G. and Teger, A. I. (1971). Reply to belovitz and finch's comments on "the risky shift in group betting". *Journal of Experimental Social Psychology*, **7**(1):84–86.

Raven, B. H. (1990). Political applications of the psychology of interpersonal influence and social power. *Political Psychology*, **11**(3):493–520.

Rogers, E. (2003). *Diffusion of Innovations*, 5th edition. Free Press, New York, NY.

Rogers, E. M., Medina, U. E., Rivera, M. A., and Wiley, C. J. (2005). Complex adaptive system and the diffusion of innovations. *Small*, **10**(3):30.

Rosenquist, J., Fowler, J., and Christakis, N. (2010a). Social network determinants of depression. *Molecular Psychiatry*, **16**(3):273–281.

Rosenquist, N., Murabito, J., Fowler, J., and Christakis, N. (2010b). The spread of alcohol consumption behavior in a large social network. *Annals of Internal Medicine*, **152**(7):426–433.

Smith, K. P. and Christakis, N. A. (2008). Social networks and health. *Annual Review of Sociology*, **34**(1):405–429.

Valente, T. W. (1996). Social network thresholds in the diffusion of innovations. *Social Networks*, **18**(1):69–89.

Valente, T. W., Watkins, S. C., Jato, M. N., Straten, A. V. D., and Tsitsol, L.-P. M. (1997). Social network associations with contraceptive use among cameroonian women in voluntary associations. *Social Science Medicine*, **45**(5):677–687.

Watts, D. (2003). *Six Degrees: The Science of a Connected Age* (1st ed.). W. W. Norton & Company, New York, NY.

Wellin, E. (1955). Water boiling in a peruvian town. In Paul, B. D., editor, *Health, Culture, and Community: Case Studies of Public Reactions to Health Programs*, pp. 71–103. Russell Sage, New York, NY.

PART III

DATA

META-NETWORKS AND RELATIONAL ALGEBRA

Intuition makes much of it; I mean by this the faculty of seeing a connection between things that in appearance are completely different; it does not fail to lead us astray quite often.

—*Andre Weil*

Learning Objectives

1. Understand source, target, and direction as it relates to networks.
2. Understand modes of data.
3. Understand meta-network ontology.
4. Use the dot product to sift common relationships in a meta-network.
5. Understand the application of "bridging" to a multimode network.

This chapter will describe methods to extract different types of networks from relational data and establish the methodology of analyzing existing relationships to determine the necessary calculations required to connect nodes of the same type through common nodes of differing types (i.e., Agents, Resources, Locations, etc.). In 1998, Krackhardt and Carley (1998) first proposed the use of matrix algebra to manipulate relational data into different network forms. We will make use of the technique in the discussion to follow.

We start by drawing the linear progression of the relationship we are trying to find from the source to the target. In Figure 7.1, the link type follows from the type of source. For instance, a dashed line indicates a source that is a Resource (i.e., car, truck, license plate, etc.) whereas a solid line represents a source that is an Agent (i.e., Ian, Tony, Avery, Marc, and Benjamin). Using this methodology, we will be able to derive distant relationships between nodes as a series of basic matrix algebra operations (Appendix A) on networks that represent those relationships. The approach used in this chapter is strictly analytical and does not attempt to factor the social theory behind the relationships, whereas Bonacich (1972) remarked

Social Network Analysis: with Applications, First Edition.
Ian A. McCulloh, Helen L. Armstrong, and Anthony N. Johnson.
© 2013 John Wiley & Sons, Inc. Published 2013 by John Wiley & Sons, Inc.

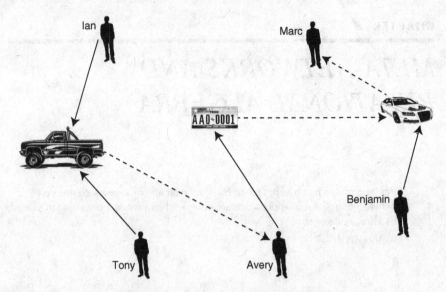

FIGURE 7.1 Multimode network of agents and resources.

that the bonds that form in a group could be conceived of as the result of "interaction potentials" possessed by each individual. On the contrary, we seek to uncover distant relationships through an objective approach based on a given relational network of observed relationships. We simply present a mathematical means of analyzing and quantifying the relational network to reveal other possible relationships that are not readily seen on the surface. This information can then be used for further analysis.

7.1 MODES OF DATA

In a single-mode network, directional network relationships are defined in terms of the relationship from the source (S) directed toward the target (T). Mode refers to the number of *node classes* present in the network. Node class is distinguished from the more familiar *node type* in that the former refers to the kind of node (i.e., Agent, Resource, Task, etc.) and the latter refers to the ways in which edges are incident with other nodes as outlined by Wasserman and Faust (1994). We use the term *single mode* to mean that there is only one node class present in the network. *Multimode* means that there is more than one node class present in the network. Meta-networks are multimode networks. Figure 7.1 is a multimode network consisting of five nodes of node class Agent and three nodes of node class Resource.

7.2 SOURCE, TARGET, DIRECTION

The directional network in Figure 7.2 shows that the S on the left has a relationship to the T on the right. This simple network is single-mode consisting of two nodes

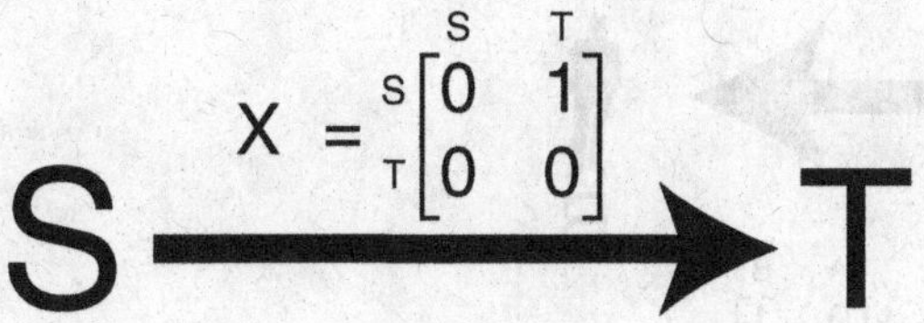

$$X = \begin{array}{c} \\ S \\ T \end{array} \begin{array}{cc} S & T \\ \left[\begin{array}{cc} 0 & 1 \\ 0 & 0 \end{array}\right] \end{array}$$

S ———▶ T

FIGURE 7.2   Source to target with matrix **X**.

and one link. The matrix **X** shows the network in matrix form. Notice that in **X** there is a relationship from the S to the T, but NOT from the T to the S.

If we want to show a directional relationship from the T to the S given this same data set, we must first create the relationship by adding a link. A directional relationship from the T on the right to the S on the left would be "reversed." The reversed link, currently, does not exist in the network **X**. To create a link that goes from the S to the T, we must take the transpose of **X** denoted as $\mathbf{X}^T$. This will show a relationship originating from the T directed toward the S. So, in terms of this methodology, whenever we want a relationship that extends in the direction from target to source, we will take the transpose of the network. Essentially, the transpose allows the relationship to trace backward from target to source. Using the transpose allows a directional relation to become bidirectional. This will be a helpful building block for later calculations.

Let us apply this methodology to a single-mode network of agents. Figure 7.3 illustrates the sources to target calculation between two agent nodes, A and B. Agent A has a relationship to Agent B. This does not mean that Agent B has a relationship to Agent A.

Example 7.1

The relationship that connects Agent A to Agent B could be "works for" (i.e., Agent A works for Agent B). Clearly, this is directional. Therefore, Agent A works for

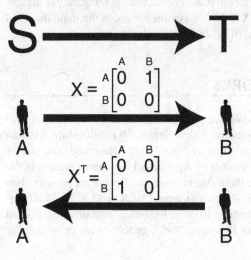

FIGURE 7.3   Source to target with matrix **X** and its transpose.

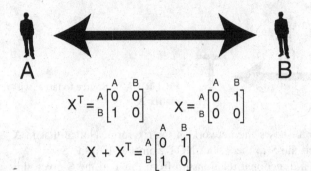

$$X^T = \begin{array}{c} A \\ B \end{array}\begin{array}{cc} A & B \\ \begin{bmatrix} 0 & 0 \\ 1 & 0 \end{bmatrix} \end{array} \qquad X = \begin{array}{c} A \\ B \end{array}\begin{array}{cc} A & B \\ \begin{bmatrix} 0 & 1 \\ 0 & 0 \end{bmatrix} \end{array}$$

$$X + X^T = \begin{array}{c} A \\ B \end{array}\begin{array}{cc} A & B \\ \begin{bmatrix} 0 & 1 \\ 1 & 0 \end{bmatrix} \end{array}$$

FIGURE 7.4 Source to target calculation for bidirectional link.

Agent B, but Agent B does not work for Agent A. This relationship is illustrated by the **X** matrix of Figure 7.3. We call this single-mode relation network **X**.

Example 7.2

Using network **X** in Figure 7.3, we can create a new network based on a different relationship, say "is the boss of," by creating a link in the opposite direction. Mathematically, this is done by the transpose of matrix **X**, namely X^T. The network X^T of Figure 7.3 shows that Agent B is the boss of Agent A. Again, this is the "reverse" relationship of **X**.

Example 7.3

There are relations that are not best described by a particular direction, but are in fact bidirectional. The network **X** showing that Agent A works for Agent B is described in the previous example. Further, from Figure 7.3, we know that X^T shows the relationship in the opposite direction; Agent B is the boss of Agent A. However, if we wanted to create the network of "works together," we need to sum the two previous matrices **X** (works for) and X^T (is the boss of). Figure 7.4 depicts this detail. Notice that the network $(X + X^T)$ is symmetric about the main diagonal of the matrix, which is characteristic of all bidirectional networks.

7.3   MULTIMODE NETWORKS

Let us apply the methodology to *multimode* networks. In Figure 7.5, we have two nodal classes or types: Agent and Resource. Let **X** denote all relationships from an Agent to a Resource. We might interpret this as Agent "uses" Resource. The size of the matrix, **X**, depends solely on the number of Agents and Resources present in the **X** network. For example, if there are three Agents and two Resources present, then the size of matrix **X** being (Agent × Resource) would be $X = (3 \times 2)$. Similarly, let **Y** denote all relationships originating from a Resource and directed toward an Agent. We might interpret this as Resource "belongs to" Agent or something similar.

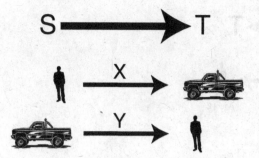

FIGURE 7.5   Two node types: agents and resources.

What if we want to show more than direct relationships? What if we want to show a directional relationship between the user of a resource, say Agent 1, and the owner of that resource, say Agent 2? We do this by taking the *dot product* of the two matrices representing the relationships **X** and **Y**. The dot product is the mathematical operation that will essentially "skip" the between-node class (i.e., the Resource) and draw a new relationship between the two agents. Recall the Goodwill Hunting problem in Section 1.4. Part two of the problem required taking the dot product of the adjacency matrix **A**. By taking the dot product twice, or **A · A · A**, the resulting adjacency matrix answers the question "find the number of walks with length three present in the graph." In other words, indirect relationships between nodes are inferred by "skipping" through the network. Likewise, indirect relationships can be inferred between nodes by multiplying matrices of relational data. But which two matrices representing networks do we use, and in which order? To determine the correct calculation, we must first look at the relationships we are attempting to combine. The following example illustrates one combination.

Example 7.4

Suppose we wanted to establish the relationship between Agent 1 and Agent 2 through Resource 1 (the truck) as depicted in Figure 7.6. This network is multi-modal by definition because we have two node classes of type Agent and Resource. This is also a meta-network because along with two node classes, there are also two networks; the **X** network consisting of a link from Agent 1 to Resource 1 and the **Y** network consisting of a link from Resource 1 to Agent 2. The **X** network contains all edges that consist of type Agent as the source and type Resource as the target. The **Y** network contains all edges that consist of type Resource as the source and type Agent as the target. Note the sizes of the two networks, **X** and **Y**. The size is based on the source and the target. Network **X** is (Agent × Resource). As there are two Agents and one Resource in the meta-network, **X** has size (2×1) and in similar fashion, **Y** has size (1×2). Our goal is to establish a link between Agent 1 and Agent 2. We require the Resource node between Agent 1 and Agent 2 to point Agent 1 to Agent 2 by surrendering its information about its own target nodes. We then must use the information to establish a link through it to the target. This process is embedded in the mathematical dot product of the two networks, **X** and **Y**. The matrix **X · Y** in Figure 7.6 is the dot product of the two networks.

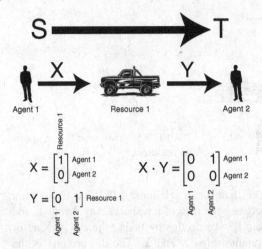

$$X = \begin{bmatrix} 1 \\ 0 \end{bmatrix} \begin{matrix} \text{Agent 1} \\ \text{Agent 2} \end{matrix} \qquad X \cdot Y = \begin{bmatrix} 0 & 1 \\ 0 & 0 \end{bmatrix} \begin{matrix} \text{Agent 1} \\ \text{Agent 2} \end{matrix}$$

$$Y = \begin{bmatrix} 0 & 1 \end{bmatrix} \text{Resource 1}$$

FIGURE 7.6 Dot product of networks X and Y: agent to resource to agent

Notice that this is an entirely new network consisting of one link, a directional link that goes from Agent 1 to Agent 2. The new link has connected Agent 1 to Agent 2 by virtue of the fact that Agent 1 is "using" a resource that Agent 2 "owns."

In Example 7.4, two network matrices, **X** and **Y**, were composed by the dot product to establish a previously hidden link between Agent 1 and Agent 2. We noted the sizes of the networks and recall from matrix algebra theory (Appendix B) that all networks cannot be composed using the dot product.

What if we want to show relationships between two users of the same resource? Figure 7.7 depicts the situation. Again, we will utilize the dot product. However, this time we have to take a closer look at the networks to compose. Previously, we defined all networks originating from an Agent (source) to a Resource (target) as the **X** network. In order to connect two users of the same resource, we need to designate one user as the source and the other user as the target. Then, we use that designation to formulate the correct strategy for uncovering the link between the two Agents.

Example 7.5

Suppose we wanted to establish the relationship between Agent 1 and Agent 2 through Resource 1 (the truck) as depicted in Figure 7.7. We first designate that Agent 1 is the source and that Agent 2 is the target. Let **X** be the previously defined (Agent $\times$ Resource) network meaning that the Agent is the source and the Resource is the target. To reach Agent 2 from Resource 1 we must use the **X** network, but this time Resource 1 is the source and Agent 2 is the target. That means we need a (Resource $\times$ Agent) relationship within network **X**. Only one operation changes rows to columns and columns to rows, or shall we say, sources to targets and targets to sources, namely the matrix algebra operator, *transpose*. Thus, the operation necessary to link Agent 1 with Agent 2 when both agents use the same resource (Agent 1 being the source and Agent 2 the target) is $\mathbf{X} \cdot \mathbf{X^T}$. Notice that

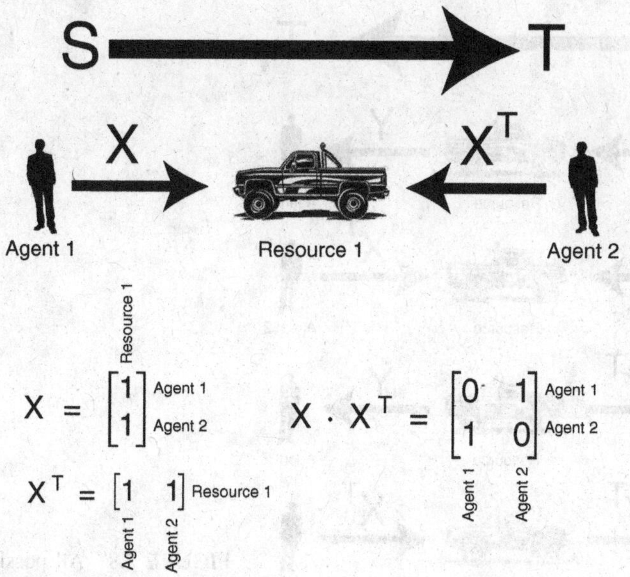

$$X = \begin{bmatrix} 1 \\ 1 \end{bmatrix} \begin{matrix} \text{Agent 1} \\ \text{Agent 2} \end{matrix} \qquad X \cdot X^T = \begin{bmatrix} 0 & 1 \\ 1 & 0 \end{bmatrix} \begin{matrix} \text{Agent 1} \\ \text{Agent 2} \end{matrix}$$

$$X^T = \begin{bmatrix} 1 & 1 \end{bmatrix} \text{Resource 1}$$

FIGURE 7.7   Dot Product of networks X and Y: both agent to resource.

the relationship is now bidirectional. This makes sense because we were finding two people joined by similar use of the same resource. Each is equally related to the other. Note that the main diagonal is set to zero because the relations are nonreflexive.

Check on Learning

Consider all possible ways in which two agents could connect to a resource. What would be the calculations used to find the underlying links that connect them together?

Answer

The four series of relationships depicted in Figure 7.8 show all of the possible combinations of relationships linking two agents through a resource:

1. Agent using a resource owned by another agent.
2. Two agents using the same resource.
3. Two agents owning the same resource.
4. Agent owning a resource used by another agent.

Remember that because these are directional relationships, we have to track them in each direction. If we want to create a social network from these two networks based on possible association through the resources, we would simply add all of

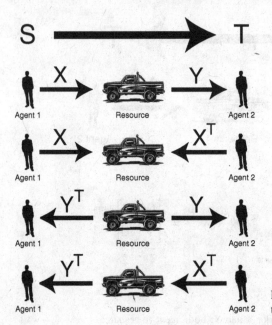

FIGURE 7.8 All possible combinations linking agents to a resource.

these relationship possibilities. The sum of these four networks created from these relationships would be equal to

$$\mathbf{XX}^T + \mathbf{XY} + \mathbf{Y}^T\mathbf{X}^T + \mathbf{Y}^T\mathbf{Y}$$

or more compactly,

$$(\mathbf{X} + \mathbf{Y}^T)(\mathbf{X}^T + \mathbf{Y})$$

7.4 BRIDGING A META-NETWORK

We have studied how to link agents based on their common relationships with the same resources. But, how do we link agents over a series of resources? In Figure 7.9, we are given a multimode network that shows agents and resources. Two agents are co-registered via license plate to a car being driven by another agent. The car meanwhile is actually owned by yet another agent. We seek to find the underlying social network that may exist between these four agents, but we have a combination that we have not encountered before. The dashed line in Figure 7.9 represents a bidirectional link between two resources. This is a very common occurrence in meta-networks. We can account for this relationship using the same technique used in the previous section. Let us define a new network matrix, say **Z**. Further, let **Z** represent the network of all links where a resource, whether source or target, connects to another resource. This network by definition must be a (Resource×Resource) network as both the source and target are the same. The size of the **Z** network in Figure 7.9 is a (2 × 2). We require that the **Z** network pass along its sources and targets, thereby creating a *bridge* for agents connected to it.

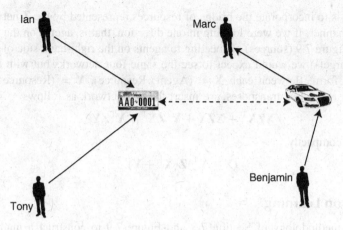

FIGURE 7.9 Bridging the gap between resources in network Z.

Example 7.6

Suppose we wanted to connect Agent Ian to Agent Marc as depicted in Figure 7.9. What intermediate steps are needed in order to make the connection? We start with the **Z** network, which represents a (Resource × Resource) network. Specifically, **Z** is the directional link between the license plate (source) and the car (target). We established which resource is the source and target by choosing which agents we want to connect. Thus, **Z** is in fact [(license plate) × (car)]. If we now wanted to represent the directional link between the car (target) and the license plate (source), we need to use the transpose of **Z**, namely $\mathbf{Z^T}$, which would be [(car) × (license plate)]. Summing the two directional networks would yield,

$$\mathbf{Z} = \begin{bmatrix} 0 & 1 \\ 0 & 0 \end{bmatrix} \mathbf{Z^T} = \begin{bmatrix} 0 & 0 \\ 1 & 0 \end{bmatrix}$$

$$(\mathbf{Z} + \mathbf{Z^T}) = \begin{bmatrix} 0 & 1 \\ 0 & 0 \end{bmatrix} + \begin{bmatrix} 0 & 0 \\ 1 & 0 \end{bmatrix} = \begin{bmatrix} 0 & 1 \\ 1 & 0 \end{bmatrix}$$

Thus, $(\mathbf{Z} + \mathbf{Z^T})$ is the intermediate step necessary to build a bidirectional bridge to connect Agent nodes Ian and Marc.

Now that we have the framework for making connections across common node classes that connect to each other as illustrated in Example 7.6, we must explore its use mathematically to make the appropriate connections. The sum of the four possible networks created previously when agents connected to each other through a single resource was

$$\mathbf{XX^T} + \mathbf{XY} + \mathbf{Y^TX^T} + \mathbf{Y^TY}$$

or more compactly,

$$(\mathbf{X} + \mathbf{Y^T})(\mathbf{X^T} + \mathbf{Y})$$

Our task is to incorporate the bridge of resources represented by the $\mathbf{Z}$ network in a similar manner. If we were looking in one direction, that is, agents on the left-hand side of Figure 7.9 (sources) connecting to agents on the right-hand side of the same figure (targets) we would expect to see the same four networks but with a slightly different form. Between each $\mathbf{X} = $ (Agent×Resource), $\mathbf{Y} = $ (Resource×Agent), and their respective transposes, we insert the ($\mathbf{Z}$) network as follows

$$\mathbf{XZX}^T + \mathbf{XZY} + \mathbf{Y}^T\mathbf{ZX}^T + \mathbf{Y}^T\mathbf{ZY}$$

or more compactly,

$$(\mathbf{X} + \mathbf{Y}^T)\mathbf{Z}(\mathbf{X}^T + \mathbf{Y})$$

Check on Learning

Use the methodology of Section 7.3 and Figure 7.9 to construct a mathematical strategy to connect agents across the license plate and the car in both directions.

Answer

Bidirectional bridging is accomplished the same way as directional bridging except between each $\mathbf{X} = $ (Agent × Resource), $\mathbf{Y} = $ (Resource × Agent), and their respective transposes, we insert the $\mathbf{Z}$ directional network as follows

$$\mathbf{XZX}^T + \mathbf{XZY} + \mathbf{Y}^T\mathbf{ZX}^T + \mathbf{Y}^T\mathbf{ZY} + \mathbf{XZ}^T\mathbf{X}^T + \mathbf{XZ}^T\mathbf{Y} + \mathbf{Y}^T\mathbf{Z}^T\mathbf{X}^T + \mathbf{Y}^T\mathbf{Z}^T\mathbf{Y}$$

or more compactly,

$$(\mathbf{X} + \mathbf{Y}^T)(\mathbf{Z} + \mathbf{Z}^T)(\mathbf{X}^T + \mathbf{Y})$$

7.5   STRENGTH OF TIES

A fundamental weakness exists with the methodology outlined in this chapter. It is very similar to the weakness Granovetter discovered in his analysis of social networks in relation to friendships (Granovetter, 1973, 1983). Namely, that most network models confine themselves implicitly with strong ties, thus limiting their applicability to small, well-defined groups. In a similar way we have done the same thing here. The analysis so far does not provide us a means to evaluate the quality of the derived connections by use of bridging. Should ties connecting agents across "class-bridges" be given the same strength as those that are connected directly? How do we account for the diameter of the meta-network in determining the strength of our links? We require a modification of our previous methodology, one that must accomplish the following three things.

1. It must detect whether or not bridging is required and perform the appropriate calculation.

2. It must weigh derived links by a scale to denote the strength of the tie.

3. It must take the number of items in a bridge into account when it scales the links.

We derive the modified scheme for inferred relationships that incorporate bridging as

$$\sum_{k=0}^{D} \alpha^k (\mathbf{X} + \mathbf{Y}^T)(\mathbf{Z} + \mathbf{Z}^T)^k (\mathbf{X}^T + \mathbf{Y})$$

where D is the diameter between the source and the target agent and α is the decay constant.

Example 7.7

Two agents are separated by two bridges of node class type Resource. Find a relational matrix algebra scheme to infer the links to connect them. Assume that the decay constant, α, is 0.23.

Using

$$\sum_{k=0}^{D} \alpha^k (\mathbf{X} + \mathbf{Y}^T)(\mathbf{Z} + \mathbf{Z}^T)^k (\mathbf{X}^T + \mathbf{Y})$$

we have

$$\sum_{k=0}^{2} (0.23)^k (\mathbf{X} + \mathbf{Y}^T)(\mathbf{Z} + \mathbf{Z}^T)^k (\mathbf{X}^T + \mathbf{Y})$$

which would be

$$(\mathbf{X} + \mathbf{Y}^T)(\mathbf{X}^T + \mathbf{Y}) + (0.23)(\mathbf{X} + \mathbf{Y}^T)(\mathbf{Z} + \mathbf{Z}^T)(\mathbf{X}^T + \mathbf{Y})$$
$$+ (0.23)^2 (\mathbf{X} + \mathbf{Y}^T)(\mathbf{Z} + \mathbf{Z}^T)(\mathbf{Z} + \mathbf{Z}^T)(\mathbf{X}^T + \mathbf{Y})$$

7.6   SUMMARY

We have established the methodology for analyzing existing relationships to determine the necessary calculations required to connect nodes of the same type through common nodes of differing types (i.e., Agents, Resources, Locations, etc.). Using a relational algebra approach does not factor any social theory behind the bonds but is merely a more descriptive aspect of the network derived from information already present. The key points discussed in this chapter are

- In a single-mode network, directional network relationships are defined in terms of the relationship from the source (S) directed toward the target (T).
- Multimode means that there is more than one node class present in the network. Meta-networks are multimode networks.
- Meta-networks are directional relationships. We have to track them in each direction. If we want to create a social network from networks based on possible association through other node classes, we simply add all of the relational networks.
- Bridging is a useful tool, but we must be careful not to over simplify.

- A fundamental weakness exists when we fail to take into account the strength of the links that are derived by bridging.

CHAPTER 7 LAB EXERCISE

Relational Algebra Laboratory Using ORA

A key strength of ORA is that it can handle networks with multiple node classes and multiple relationships. In fact, the matrix algebra conducted in this chapter is easily done in ORA. However, understanding how to combine relations (the relational algebra) cannot be done by a computer. It requires an analyst with a sound understanding of relational algebra. This laboratory will provide practical, hands-on experience in using ORA to uncover implicit connections using relational algebra.

Step 1. Create three different node classes for Agents, Locations, and Resources.

Step 2. Create networks (Agent × Location), (Agent × Resources), (Resource × Agent), (Location × Location), and (Resource × Resource) using Tables B.10, B.11, B.12, B.13, B.14 as outlined in Appendix B.

When visualized you should have a multimodal meta-network similar to the one below.

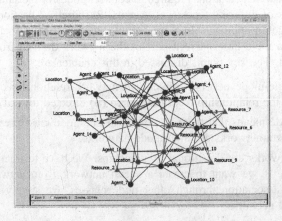

Step 3. What are the different social networks that you can extract from this data? For example, can you develop a social network based on shared resources?

In this laboratory, there are two networks that have links connecting people and resources. There is an (Agent × Resource) network and there is a (Resource × Agent) network. In the first network, the agents are providing some kind of resource. In the second network, the agents are using some resource. We may want to connect people based on shared contributions to a resource. If the (Agent × Resource) network is $\mathbf{X}$, then the desired network would be $\mathbf{X}\mathbf{X}^T$. In this network agents are connected if they both contribute to the same resource. If you want to connect people who draw from the same resource, and the (Resource × Agent) network is $\mathbf{Y}$, you would need to calculate $\mathbf{Y}^T\mathbf{Y}$. If you want to connect people who contribute to a resource to those using the resource, then you would calculate $\mathbf{X}\mathbf{Y}$. You may also want to calculate the reverse relationship $\mathbf{Y}^T\mathbf{X}^T$.

Step 4. In order to calculate these quantities in ORA, you must first obtain the transposes of the (Agent × Resource) and the (Resource × Agent) matrices. ORA's Matrix Algebra dialog box contains a transpose function to accomplish the task. You can locate it by choosing *Data Management → Matrix Algebra . . .*, then select the transpose operation. It is important to note that the transpose of the (Agent × Resource) matrix is not the same as the (Resource × Agent) matrix. Only their sizes are the same.

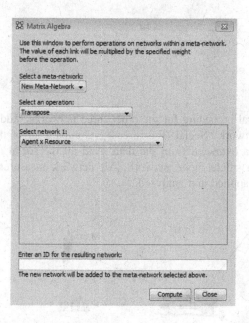

Another way to calculate the transpose is to right click on the (Agent × Resource) network in the Meta-Network Manager pane and select *transpose*. Use the default label. Right click on the (Resource × Agent) network and select the transpose. To calculate $\mathbf{X}\mathbf{X}^T$, once again, select *Data Management* and then *Matrix Algebra . . .*

from the pull-down menu. Select the operation "Multiply Networks." The first network is the left network in this case (Agent × Resource). The second network will allow only networks where the rows are resources. This is because ORA will force you to use proper dimensional analysis before it will conduct an operation. Select the "Agent × Resource transposed" from the drop down menu. Name the new network "XXt (Resources)."

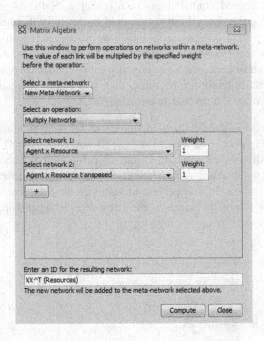

Repeat this general procedure for the other three networks. Add them all together and name the network "Social (Resources)." Compare all of the different social networks that you extracted from the data. Add all the social networks together. You now have a single mode network. All network measures for single mode networks can be applied and analyzed.

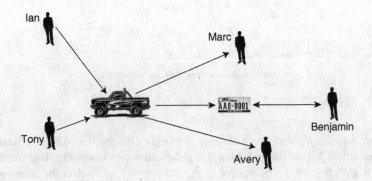

EXERCISES

Use the meta-network discussed earlier to complete the following exercises.

7.1. Write the adjacency matrices for the following networks.

    1. $\mathbf{X} =$ Agent $\times$ Resource

    2. $\mathbf{Y} =$ Resource $\times$ Agent

    3. $\mathbf{Z} =$ Resource $\times$ Resource

7.2. Use matrix algebra to calculate

$$\mathbf{X}\mathbf{X}^{T}, \mathbf{X}\mathbf{Y}, \mathbf{Y}^{T}\mathbf{X}^{T}, \mathbf{Y}^{T}\mathbf{Y}$$

7.3. How many agent $\times$ agent links emerge in each of the following networks?

    1. $\mathbf{X}\mathbf{X}^{T}$

    2. $\mathbf{X}\mathbf{Y}$

    3. $\mathbf{Y}^{T}\mathbf{X}^{T}$

    4. $\mathbf{Y}^{T}\mathbf{Y}$

7.4. Determine what relational algebra is required to connect Agent Ian to Agent Benjamin. (*Hint: You must use the* $\mathbf{Z}$ *matrix*)

7.5. What relational algebra is required to connect Agent Ian to Agents Avery and Tony? What other agents become connected as a result?

7.6. What relational algebra is required to connect Agents Avery and Marc?

REFERENCES

Bonacich, P. (1972). Factoring and weighting approaches to clique identification. *Journal of Mathematical Sociology*, **2**(1):113–120.

Granovetter, M. (1973). The strength of weak ties. *American Journal of Sociology*, **78**(6):1360–1380.

Granovetter, M. (1983). The strength of weak ties: A network theory revisited. *Sociological Theory*, 1:201–233.

Krackhardt, D. and Carley, K. (1998). A pcans model of structure in organization. *Proceedings of the International Symposium on Command and Control Research and Technology*, pp. 113–119.

Wasserman, S. and Faust, K. (1994). *Social Network Analysis: Methods and Applications*. New York, NY: Cambridge University Press, pp. 39–40.

CHAPTER 8

SOURCES OF DATA

It is a capital mistake to theorize before one has data.

—*Sir Arthur Conan Doyle*

Learning Objectives

1. Understand modes of data.
2. Understand boundary specification (population, sampling, realist, nominalist).
3. Understand snowballing.
4. Know the collection methods for social network data.
5. Explain concepts: roster versus free recall; free versus fixed choice; rating versus ranking.
6. Understand issues surrounding anonymity.

Methods of data collection and the way we represent data have a large influence on the results we will obtain from using social network analysis tools and methods. Research data is collected to help us understand a situation under study and to test hypotheses about what is happening within that context. Careful consideration must be given to the matching of data types and formats with the research questions. We need to be clear about what we are trying to achieve in collecting data and analyzing a particular network, and the data we need to collect to answer the underlying questions. Once we are clear about the objectives, we can design the most appropriate approach to data collection and its coding in the network.

8.1 NETWORK SAMPLING

There are two types of data in social network analysis: structural and composition. Structural data describes pairs of dyads or triples of triads. It also includes information such as topology. Composition data consists of attributes and measures on the

Social Network Analysis: with Applications, First Edition.
Ian A. McCulloh, Helen L. Armstrong, and Anthony N. Johnson.
© 2013 John Wiley & Sons, Inc. Published 2013 by John Wiley & Sons, Inc.

189

individuals in question. For example, measuring the age, gender, race, socioeconomic status, and height of individuals are all forms of composition data. Kinship, affiliation in organizations, and friendship are all forms of structural data.

Nodes in a network do not have to represent the same thing. Some nodes can be people; other nodes can be locations, organizations, beliefs, events, tasks, resources, etc. If a network contains multiple types of nodes it is multi-modal. If for example, nodes were used to represent people and locations, the network would have two modes. If the network also had events, the network would have three modes, and so on. Therefore, the number of *modes* refers to the number of different types of nodes in the network.

An important concept in collecting network data is the *boundary specification* (Wasserman and Faust, 1994). This refers to a problem of determining whom to include in a social network study. There are two classic approaches: the realist and the nominalist. The realist approach identifies the group based on the member-defined identity. The New York Giants have defined members that belong to the team. An Army platoon has members that definitely belong to that group. The nominalist approach defines a group for some scientific objective. For example, if someone wanted to study the interactions between psychologists, they might define a group as the individuals who have published papers in the *American Journal of Psychology*. While there are certainly people who might consider themselves psychologists who have not published in this journal, the group is convenient for the research question under investigation.

For many social network studies, groups are small enough to collect data on all individuals in the network. In some cases, however, the group is too large. The population refers to all individuals in the group being studied. A sample refers to a random subset of these individuals. A sample in a network sense is a bit different than in a traditional statistics sense, because we are most interested in preserving the network structure that is representative of the population. For example, in a social network study conducted at the US Military Academy to understand a cadet military chain of command, all individuals in one cadet regiment were selected. There were not enough resources to collect data on all cadet leaders within the Academy, therefore only a limited number of cadets could be studied. If these cadets were chosen at random from across the Academy, it might be reasonable to assume that a commander in first regiment would not communicate with a commander in fourth regiment due to distance and the difference in the chain of command. Therefore, the entire chain of command in the fourth regiment is chosen, so that this regiment might be considered a representative random sample of the structure of a chain of command at the Academy.

Another method of collecting data on large groups is snowballing, also known as *respondent-driven sampling* (Goodman, 1961; Salganik and Heckathorn, 2004). In *snowballing* an agent of interest is selected and all of the other agents connected to this first agent are selected. These nodes are called the *first-order connections*. Next, the agents connected to all of the first-order connections are identified. These nodes are called *second-order connections*. This type of data collection is often useful to explore the social network space surrounding some agent of interest.

Several concerns arise in snowball sampling. One concern is that the size of the network can grow quite rapidly as this process continues. Another concern is that individuals not relevant to the social group under research might be included. For example, if a criminal investigator was building a gang network based on warrant-supported phone tapping, there may exist nodes in the network that are not affiliated with the gang. In addition to gang member to member communication, they may also communicate with family members, non-gang affiliated friends, and others. A common approach to handling this problem is to use the "2.5 degrees" rule-of-thumb. This involves snowballing three degrees from the initial seed node, or third-order connections. Then remove all pendant nodes (nodes with only one link into the network). In a sense, this screens out individuals who are not likely to be affiliated with the group under study.

Check on Learning

True or False. The rule of thumb for snowballing is to include first-order connections, second-order connections, and third-order connections, then remove any nodes with only one link into the network.

Answer

True. This method ensures that the sample size is not overwhelming and limits the scope to those individuals who are most likely to be of interest.

8.2   MEASURING LINKS

There are many ways in which nodes may be linked together. A person may have a set of individuals they consider friends in a friendship network. There may be other people they go to for advice. Other people might be kin relatives. Some people might be affiliated by common membership in organizations. There are six categories of relationships that we may wish to measure in a network.

Individual Evaluations: These are relations defined by the judgment of individual actors, such as friendship, trust, and respect. These relationships are most common in social network analysis.

Transaction: This relationship involves the transfer of some material resource, such as lending money and buying. Once the resource is passed from one actor to the next, the original actor no longer has that resource. This type of link is used primarily as a proxy for some other relationship. In order for one actor to give a resource to another actor, they must have some other type of relationship ranging from acquaintance to stronger affinity.

Transfer: This relationship involves the transfer of nonmaterial resources. Unlike transaction, both agents may possess the resource after transfer, such as communication, knowledge, or sexually transmitted disease.

Affiliation: Some agents may share affiliation with the same organization, live in the same house, or attend the same school with each other.

Formal: This relationship is formally established and does not require the opinion of the node. For example, a chain of command follows this relationship.

Kinship: Similar to the formal relationship, but with more defined meaning (i.e., sibling, parental, tribal).

All relationships can tell us something about the network and the how people within it associate or show common ties. Relationships may be binary (0 or 1), negative or positive (− or +), carry a specific value (exact number such as dollars, speed or distance, etc.). The absence of a link may also be significant in some networks, such as linking evidence to suspects in a criminal network. The social network researcher must determine the most appropriate way to design and measure the value of a relationship and how it should be defined, based on the research question.

There are important conceptual and theoretical differences between positive, negative and null ties. In a positive tie network, a node high in betweenness centrality is in a position to broker information and resources between actors in the network. This allows that highly central node to hold informal power. This is not necessarily true for a null network or negative tie network. Two nodes in the positive tie network with no link between them may represent a null tie, where there is no relationship between the actors. It could also represent a negative tie, where the actors dislike each other. In a negative tie network, betweenness centrality (and other node level measures) loses its conventional interpretation.

Example 8.1

Consider a network, where the links represent dislike instead of liking. A node on the path of dislike derives no informal power, positional advantage, or positional disadvantage. Keep in mind that in the negative tie network, two nodes that do not have a dyadic link may either like each other, or have no relationship at all.

Treatment of negative and null ties is an area of active research. For our purpose, we present the issue as a data collection concern. When collecting social network data, it is important to keep in mind that there are many ways in which two nodes might be related. Sometimes, the absence of relationship (null tie) is the interesting network to explore. Other times it is the positive tie network, negative tie network, or another type of relation completely. Two nodes that consider themselves friends do not necessarily trust one another. Two nodes that communicate for work related issues may or may not engage in gossip or discussions about their personal life. Data collection design must carefully consider the relationships of interest and determine the value network centrality measures may or may not provide in light of those relationships.

There are several ways to collect data regarding relationships in social networks. The key methods we discuss are questionnaires, interviews, observation, archival records, and e-mail.

Questionnaires are the most common method of collecting social network data. Subjects may be asked about many different relationships. There are several key experimental design considerations for questionnaires.

The first consideration is *roster versus free recall*. The roster design will provide the subject with a list of possible choices. The free recall method on the other hand requires the subject to remember who his friends are without any prompting from the questionnaire. For example, if you ask a subject to name their close friends, the roster method may provide them a list of people in their workplace, of whom they can choose others to link to. In this method, they are not able to select individuals that are not on the list. Research has shown that people can often forget close friends if they are not prompted and that free recall is more affected by how recently a subject has interacted with others than by the closeness of a relationship. Humans also tend to categorize acquaintances. Therefore, they are more likely to recall certain subgroups and leave out others based on their individual categorization. Some feel, however, that the roster method limits the potential relationships that a subject can name and therefore biases the study.

Example 8.2

McCulloh and Geraci (n.p.) conducted a study of post-traumatic stress disorder (PTSD) which included social network data on a US Army Infantry Brigade consisting of almost 1000 soldiers. Data was collected prior to the brigade's deployment to Afghanistan, 2 months into their deployment, and after they returned from deployment. The first questionnaire asked respondents to "list their close friends within the [military] unit." Example responses were "Bubba," "My wife," "Big Joe," "Smitty," and other unusable responses. They were able to identify usable data from only 2 out of 12 companies in the brigade.

There were two challenges that the researchers faced. The first was the impracticality of providing respondents with a roster consisting of 1000 names on it. The second was the concern associated with prompting the respondent. Qualitatively, they determined that most relationships between soldiers were within the company level of the organization. Companies consisted of between 50 and 170 soldiers, depending upon the manning and purpose of the company. During the second iteration of data collection, respondents were provided company rosters. They were asked to place a check next to the individuals they considered to be a close friend. Usable data was collected on all companies surveyed. However, some respondents that completed surveys were not included on the roster, because they had been recently transferred into the unit after the roster was made. This provided another source of error in the study.

Roster and free recall both have strengths and limitations. A design decision must include considerations such as feasibility, size of the network, accuracy of the roster, among other issues. The third iteration of data collection used the same protocol as the second iteration.

Another consideration is *free versus fixed choice*. Free choice allows the subject to select as many potential others to link to. The fixed choice design, on the other hand, limits the number of others to some defined number. If you wanted to know close friends, the free design would allow the subject to name 3 or 30. The fixed choice design sets a number; for example 5; and then the subject must name exactly five friends. If they only have three close friends, they must arbitrarily choose two more. If they have six close friends, they must leave one out.

Yet another consideration is *ratings versus rankings*. Rankings require a subject to list the ranking of the strength of ties. For example, if you were investigating friendship with the ranking approach, you would ask the subject to rank order his friends from one to however many friends he has. The ratings approach measures the strength of relationship, allowing for ties. For example, if you investigate friendship with the rating approach, you would ask the subject to rate his friendship on a scale of 1 to 10. Some argue that if possible, you should always attempt to obtain rankings over ratings when possible, because there is higher resolution in rankings. However, if an individual has a few close friends, more regular friends, and some acquaintances, the ranking difference between a close friend and a regular friend would appear the same as the difference between close friends. Ratings, on the other hand, may not align with the culturally defined categories intended for collection.

Usually when we use a ratings approach, we prefer the Likert style scale, which is a five-point (0–4, 1–5) or seven-point scale (0–6, or 1–7). This scale allows a midpoint for exact ambivalence, extremes, but requires the user to push a little more to either the middle or extreme for other ratings. Research has found that 3- and 10-point scales can be more biased by people completing the survey than the five- or seven-point scales. However, it is often wise to use a scale that has been established in the literature in order to establish credibility in your research.

It is also very important to note that people can assess friendship differently. Strength of friendship of 6 to one person may seem like a 5 to someone else. It is wise to ask an objective question where possible for ratings. For example, instead of asking to rate their friendship on a scale of 1–7 where 7 is a close friend and 1 is an enemy, it may be more wise to ask, do you avoid this person, are you acquainted with this person, do you like this person, do you associate outside of work, do you go to each others house to socialize, do you go on vacation together, are you in an intimate relationship with this person. This makes the assessment of relationship more objective and consistent between nodes and dyads.

A different approach to questionnaires is *interview*. Interviews are used when questionnaires are impractical or the investigator lacks enough knowledge about the subject group to design appropriate questions. In some cases, interviews are used on the front end to provide insight into the nature and type of relationships and composition data that are important for the group under study. Interviewing is probably the most challenging method of data collection for the researcher. Effective interviewing requires the interviewer to gain rapport, convince people to open up, record accurate notes, begin and end the interview. These skills must be developed with practice.

Another data collection approach is *direct observation*. Observers need to be precise and consistent with their identification of relationships. One study at a military training event required the observer to record the number of statements/commands sent between members in a platoon chain of command during convoy operations. Early missions had very little communication where density was less than 0.3. Later missions had high communication with density greater than

0.8. The density of this network was highly correlated with the notional casualties incurred in the mission.

Archival records are another source of data. Many relationships can be defined for a set of records. A classic example is author co-citation networks. This establishes a link between individuals who have coauthored papers together. One famous mathematician, Paul Erdos published a large number of academic papers. Mathematicians like to track how many shared coauthors they are away from Paul Erdos. This is called an *Erdos number*. The lower the Erdos number, the more prestigious the mathematician.

A specific form of archival records is *e-mail data*. Most e-mail exchange servers maintain header information from e-mail traffic. This data consists of the TO, FROM, CC, BCC, SUBJECT, E-MAIL MESSAGE ID, and the date and time stamp. The header information provides necessary data to construct an e-mail communication network. E-mail communication does not necessarily reveal friendship, trust, or advice seeking. The volume of e-mail activity is usually more a function of an individual's role within the organization and their personal e-mail habits. Some people will e-mail the person sitting next to them. Others will walk to another room down a hallway to speak to someone face-to-face. However, the presence of an e-mail link between people demonstrates that there is at least some level of relationship between actors. Attribute defined subgroup analysis can help average actor e-mail behavior across groups and provide subgroup to subgroup communication behavior.

8.3   DATA QUALITY

When we discuss data, there are several concepts that refer to how good the data is (Wasserman and Faust, 1994).

Accuracy refers to the reported information being actually correct. It is well established that if you ask someone to tell you the people they spoke with last week, they will usually forget someone. They may also report someone who they spoke with 2 weeks ago, but not last week. There are several reasons that people report inaccurate information. Bernard (2013) offers four key reasons.

People who participate in a survey or interview have a stake in the process and will attempt to answer all of the questions. They may or may not understand the question, but will still try to provide an answer. Some respondents will provide responses that they think the researcher is looking for whether or not it is accurate.

It is easier for people to remember some things more than others. Memory is fragile. It is easier to remember recent events, rare events, and significant events. Common behavior over a long period of time causes people to estimate their behavior rather than report their actually memory of the events.

Social network data collection is a social encounter. Individuals will tend to respond in ways that offer them a perceived advantage. For example, boys tend to exaggerate reports of their sexual experience, while girls tend to minimize reports.

People use rules of inference because they are unable to remember actual events. Freeman et al. (1987) asked people to list those who attended a colloquium at their university. People tended to report those who usually attended the colloquium, rather than those who actually attended.

Some forms of data collection are more accurate than others. Questionnaires and interviews rely on human memory for the data and are often less accurate than direct observation, archival records, or e-mail data. It is possible, however, that an observer failed to notice some interaction or the context in which interaction may have occurred. Just because data is more accurate, does not mean it is of better quality.

Validity refers to whether the measurement captures what it is supposed to. For example, using e-mail data to measure a friendship network may not be valid. It is likely that a person may e-mail coworkers, receive junk mail, and have links to others who are not necessarily friends. Asking the person would reveal real friends from other non-friend links. There is often a trade-off between accuracy and validity.

Reliability refers to consistency in the measurement. A measure is reliable if the same measurement can be recorded multiple times and the same results are obtained.

Finally, *measurement error* occurs when there is a discrepancy between a true relationship and the value that is recorded.

Several techniques can be used to improve the quality of data collection. *Cued Recall* is a process where respondents review records such as phone logs, letters, or membership rosters to cue their memory. This approach has not been shown to create 100% accuracy, but it has been demonstrated to improve a respondent's accuracy.

Landmarks is a method where people are provided a mental landmark such as the birth of a child, graduation, or major news event and ask people to report activity since the event. Offering a respondent multiple landmarks may help reduce their tendency to rely on rules of inference for responses.

Restricted Time is a technique where the recall period is kept short. The short time period allows people to focus and provide more accurate responses.

8.4 ADDITIONAL ETHNOGRAPHIC DATA COLLECTION METHODS

Cultural domain analysis is a method to map the cultural domains of social groups. Cultural domains are the classification schemes that people use. For example, if you were to ask a group of people on the beach to identify which of the three animals is not like the others: DOLPHIN, SHARK, DOG; they are likely to say "DOG." The other two live in the water. If you asked the same question of a group of biologists, they might be more likely to respond, "SHARK," because the other two are mammals and the shark is a fish. Neither response is right or wrong. They are simply different cultural domains by which people classify things.

Cultural domains are determined by asking respondents a question requiring a free list response. For example, Casey and McCulloh (2012) asked people in public markets in Mosul, Iraq what are characteristics of a hero? Respondents were then encouraged to list as many responses as they could and were prompted, "what else?" The earlier a person listed a response and the more common the response is across respondents, the more salient the response.

Networks can be created from cultural domains in different ways. Social networks can be constructed based on actor similarity in responses. Two-mode networks can be constructed where the respondents are agents, and the responses are the second mode. Relational algebra can fold the two-mode network into a social network based on a salience-weighted threshold or a (response×response) network can be created based on common respondents. Subgroup analysis of these networks can reveal differing cultural groupings.

Name Generators are used to construct ego networks. A respondent is asked questions such as "who would you ask for a loan," "who would you ask for work related advice," and "who would you ask about employment opportunities." Name generators are used to build networks, where the population is ill-defined, with no convenient boundary specification. Name generators are often used in conjunction with snowball sampling to identify social groups.

A challenging method of collecting network data comes from text analysis. Terrorism experts will read through intelligence reports, summaries, news media, and more to extract important relationships between terrorists. Financial experts will listen to the speeches of central bankers and financial news reports to extract meta-networks pertaining to finance. Criminal investigators will extract meta-networks of evidence from witness statements and other forms of text.

Here, we present a systematic approach for extracting meta-networks from text based upon the Caleb methodology frequently used in intelligence analysis. The first step is to determine the context or application area for the text extraction that is most relevant to the research objectives. Are you building a terrorist network, financial network, crime scene network, or something completely different.

Second you must decide on your meta-network ontology. The basic ontology consists of Agents, Knowledge, Tasks, Resources, and Organizations. If you are investigating terrorism, you may want to include events and locations. If you are investigating financial networks, you may wish to include stocks. If you are investigating crime, you may wish to include evidence. There is no set ontology that the analyst must use. Rather, it is tailored to the particular application.

Third, you must decide on what meta-relations that will be explored. Are you looking at Agent × Agent networks? Agent × Organization networks? Task × Resources networks? You do not have to consider all ordered pairs of network relationships and you can include multiple types of relationships within a meta-relation. For example, in the Agent × Agent meta-network, you can investigate friendship networks, trust networks, kinship networks, and many more. All of these networks may include the same individuals. The investigator defines the relationships or links.

TABLE 8.1 Caleb Methodology Table

Source Node ID	Source Node Name	Target Node ID	Target Node Name	Strength of Relation	Statement	Source	Time of Incident	Time of Record
A102	Joe	T23	Drive	3.0	Joe drove the truck	Doc #43	2:00pm 2/14/09	2/18/09
A103	Frank	T23	Drive	−2.0	Frank cannot drive	Doc #44	2/16/09	2/17/09
...	...	...	...	...	...	...	...	...

Fourth, once these decisions are made, the analyst can begin to explore the document and record structural data ready for analyses. For each set of network relations the information displayed in Table 8.1 should be recorded. The analyst may create a separate table for each set of relationships. The source and target node ID columns provide the nodes for the network. The strength of relation provides potential weight for the links. The statement and source provide validation of the relationship. The time columns allow data to be segregated into time periods for future longitudinal analysis.

8.5 ANONYMITY ISSUES

In defining the dataset required for the analyses the boundaries must be defined, that is, what data do we want to collect from whom, and do we need to ensure anonymity of the participants? Once we have identified who we need to collect from we then need to determine how many participants, and what level of detail is needed in the data. Do we need to include the entire network or will a sample give sufficient rigor to our findings? If a sample is to be collected, what size sample are we seeking and from where will we draw this sample to ensure no bias in selection of participants? In some cases informed consent and/or anonymity will be required, particularly where sensitive information is being analyzed, in such cases as networks of drug users, contagious diseases with anti-social connotations or suspects of crime networks. The process of replacing an identity with a code is straightforward, however those who know the network well will be able to reverse engineer the coding and identify or infer identities just from their positions in the network or relationships. Kadushin (2005) observes that data collected on individuals or social units is the very point of the research, not incidental to it. Borgatti and Molina (2003, 2005) also highlight the importance of identities in the building of sociograms where named connections are important to the construction of the network. Borgatti suggests the researcher be cognizant of who is to see the data and what the data will be used for, and consider whether the participants would be strengthened or weakened by the results of the analysis, observing that the stakes are higher in the managerial context than in academia. Borgatti and Molina (2003) recommend two principles to guide the researcher: avoiding harm to innocents and providing value to participants.

However, in most cases social network analysis provides a wealth of information not only about the people by their attributes but also their position in the network, the network structure and the relationships between those in the network. In some cases meaning cannot be derived unless the participants are identified. Anonymity is suitable where the research is focused on the structure and dynamics of the network, however, where participants are requested to identify people they trust, report to, or with whom they collaborate, for example, then anonymity will generate results of limited value.

8.6 SUMMARY

Our ability to answer research questions with confidence depends directly upon the quality of the data gathered and the application of appropriate analyses to that data. Deciding on what data to gather, its format, and the methods of collection will affect the accuracy, validity, and reliability of the data. Bias in data collection and collation should be minimized to ensure valid deductions from our findings. The key points discussed in this chapter are

- There are two types of social network analysis data: structural data comprising of pairs of dyads and triples of triads, and composition data comprising of attributes and measures relating to the individuals.
- When defining boundaries in social network analysis it is important to preserve the network structure, so care must be taken when deciding upon criteria for inclusion in data gathering. In large networks where all individuals cannot be included the criteria for inclusion in the chosen sample must reflect the best data set available to achieve the research objectives.
- Snowballing should set the boundary for sample inclusion to no more than three order connections from the initial node of interest.
- Relationships in networks fall into six main types: individual evaluations (value placed on relationship by the individual), transaction (resource is passed and source no longer possess the resource), transfer (resource is passed but both agents possess the resource), affiliation, formal, and kinship.
- Link values can be binary (0 or 1), negative/positive (+ or −), or carry a specified value, and nonexistence of links may also be significant.
- Data in questionnaires can take several formats: roster providing possible choices, free recall requires the respondent to think of the answer, free choice permits the respondent to select the number of inclusions, fixed choice sets limit to the number of inclusions, ranking requires the respondent to rank their choice in order and rating enables the respondent to scale the strength of the relationship.

- Data quality considerations include accuracy, validity, and reliability, measurement errors should be minimized by the use of cued recall, landmarks, and restricted recall time periods.
- Cultural domain analysis and name generators can be used to gather data to populate a network.
- Extracting meta-networks from text can be achieved through the following steps: determine the context and objectives, decide upon on the meta-network ontology, identify the relevant meta-relations to be explored and the associated networks, collect data and populate in tables of relationships ready for analyses.
- Determine whether the data collected requires anonymity, considering most social network analysis provides valuable information about the position and relationships of named entities.

EXERCISES

8.1. Ask 10 people to list their friends. Keep encouraging them to list more. When finished, ask them to categorize their friends. In other words, how do they know the people? What kind of friend are they? What do they do together? Then, go back through the list and ask them to categorize each friend. In general, categories should be clumped together. Why is this? What can this tell you about possible bias in free recall?

8.2. Are there any individuals in your 10 respondents that are listed in other respondent's friend lists? If so, are they reciprocally listed? What technique involves using the first respondent to develop the list of subsequent respondents? What are strengths and weaknesses of this method?

8.3. There are six categories of relationships presented in this chapter. Compare, contrast, and provide examples of a relationship for each category.

8.4. Discuss the data quality issues with the following methods of social network data collection:

1. e-mail;
2. interview faculty of an elementary school;
3. questionnaire to collect friendship and respect relations in a military unit of 100 soldiers;
4. observe student interaction in a second grade classroom.

REFERENCES

Borgatti, S. and Molina, J. (2003). Ethical and strategic issues in organizational social network analysis. *The Journal of Applied Behavioral Science*, **39**(3):337–349.

Borgatti, S. and Molina, J.-L. (2005). Toward ethical guidelines for network research in organizations. *Social Networks*, **27**(2):107–117.

Casey, K. and McCulloh, I. (2012). Hts support to information operations: An example of integrating hts into coin operations. *Military Intelligence*, **37**(4):28–32.

Freeman, L. C., Romney, A. K., and Freeman, S. C. (1987). Cognitive structure and informant accuracy. *American Anthropologist*, **89**:311–325.

Goodman, L. A. (1961). Snowball sampling. *Annals of Mathematical Statistics*, **32**(1):148–170.

Kadushin, C. (2005). Who benefits from network analysis: ethics of social network research. *Social Networks*, **27**(2):139–153.

Salganik, M. and Heckathorn, D. (2004). Sampling and estimation in hidden populations using respondent-driven sampling. *Sociological Methodology*, **34**(1):193–240.

Wasserman, S. and Faust, K. (1994). *Social Network Analysis: Methods and Applications*. New York, NY: Cambridge University Press, pp. 30–33.

PART IV

ORGANIZATIONAL RISK

CHAPTER *9*

ORGANIZATIONAL RISK

> Irrational action is the logical result of poorly defined performance measures.
> —*Eli Goldratt*

Learning Objectives

1. Explain the concept of risk in terms of social networks.
2. Describe the threats to social networks.
3. Explain the measures used to investigate network risk.
4. Understand potential risks in organizational meta-networks.
5. Understand the process of identifying risk in an organizational network.

Every decision we make in life is based upon an unconscious evaluation of the risks and benefits that could result from our decision. Even the decision to cross a road involves a subconscious evaluation of consequences—and the impact will be different depending upon how heavy the traffic flow is, quality of sight of oncoming vehicles, whether we are crossing at traffic lights or a pedestrian crossing, the weather conditions, and so forth. We would normally make sure we are well informed about the consequences of decisions that hold high risk before we commit to a decision. According to the Oxford English Dictionary (http://oxforddictionaries.com/definition/risk) the term *risk* originates from the early to mid-1600s and is defined as "the possibility that something unpleasant or unwelcome will happen," "a situation involving exposure to danger," and "expose (someone or something valued) to danger, harm, or loss."

9.1 WHAT IS RISK?

The ISO 31000 standard on Risk Management (ISO 31000, 2009) defines *risk* as "effect of uncertainty on objectives" where an effect is a deviation from the expected that can be either positive or negative, and objectives can relate to financial, health and safety, and environmental goals and these can apply at strategic,

Social Network Analysis: with Applications, First Edition.
Ian A. McCulloh, Helen L. Armstrong, and Anthony N. Johnson.
© 2013 John Wiley & Sons, Inc. Published 2013 by John Wiley & Sons, Inc.

organization-wide, project, product, and process levels. The standard also describes uncertainty as "the state, even partial, of deficiency of information related to, understanding or knowledge of an event, its consequence, or likelihood" (ISO Guide 73:2009, Definition 1.1).

Risk analysis is a fundamental management tool. It enables us to make more informed decisions that hopefully result in better outcomes than decisions based upon no analysis. The determination of risk can provide more peace of mind for the decision maker as the more information we have regarding a situation, that is, situational awareness, the more confident we can be about the outcome of our decisions. Risk comes in many forms and in the context of an organization is concerned with the nonreaching of objectives.

Vulnerabilities are shortcomings that leave the organization open to exposure. The fact that you have no firewall installed on your computer network is vulnerability. *Threats* are unwanted happenings that could cause loss, events that take advantage of vulnerabilities, that is, the threat that your computer system may be hacked and your credit card details stolen. *Risk* is the probability that a threat may happen and in the case of having no firewall installed the probability that your system will be hacked will be pretty high. Risk is also used frequently to mean threat. The impact of the threat is usually noted in dollar terms by estimating the losses that would result if a threat were to occur. The impact of your system being hacked could be the loss of all your data and the cost of retrieving it, or loss of sales and/or customer confidence if your system needs to be constantly available or even loss of life if your system is an emergency response or life dependent system.

The ISO risk management standard (ISO 2009) also describes the internal and external factors that should be considered when managing risk. The external environment in which the organization seeks to achieve its objectives can include:

- the cultural, social, political, legal, regulatory, financial, technological, economic, natural, and competitive environment, whether international, national, regional, or local;
- key drivers and trends having impact on the objectives of the organization; and
- relationships with, and perceptions and values of, external stakeholders.

The internal environment in which the organization seeks to achieve its objectives can include

- governance, organizational structure, roles, and accountabilities;
- policies, objectives, and the strategies that are in place to achieve them;
- the capabilities, understood in terms of resources and knowledge (e.g., capital, time, people, processes, systems, and technologies);
- information systems, information flows and decision-making processes (both formal and informal);
- relationships with, and perceptions and values of, internal stakeholders;
- the organization's culture;

- standards, guidelines, and models adopted by the organization;
- the form and extent of contractual relationships.

All these factors can have an effect on the outcomes of our decisions, and it is wise to consider the relevance of such factors in our evaluations of risk.

In the case of networks risk can relate to either nonattainment of the objectives of the system under study via imbalance within, or threats to, the network structure or dynamics. Traditional risk analysis relates to the nonattainment of objectives by identifying assets and threats, quantifying or qualifying the impact of occurrence of the threat or losses, and the probability of occurrence. Threats in traditional risk analysis relate directly to the objectives—customer satisfaction, profits, achievement of goals, etc. While also related directly to achievement of objectives, threats to networks take a slightly different form.

9.2   MEASURES OF CENTRALITY AND RISK

Social network analysis provides us with extensive measurements for analysis of an agent's position in the network and their influence based upon their centrality. Position in the network, however, can be used to positively support cohesion in the network and jointly support organizational goals or alternatively can be a threat to the ongoing operation of the network. It would be easy to assume that the person with the highest degree centrality is the most influential; however, this is not always the case. The agent who has the highest degree centrality and is most connected to others may not be the most critical. Many influential nodes have connections only to other nodes in their immediate cluster or clique, connecting solely to those nodes that are already connected to each other. People have access to different information and have varying influences across the network and we need to identify who is the right source of information, who is crucial to the flow of information and has access to resources and knowledge, and who is able to influence others in the network.

Social network centrality measures provide insight into key people within the organization (Chapter 2). Network centralization measures provide an indication of how much a network is dominated by a few highly central nodes (Chapter 3). Individuals with high betweenness centrality have influence and power in the network. If only a few individuals show high betweenness then power is held by these few, or they are responsible for too many key areas and are candidates for high stress. Such powerful individuals are targets to remove from the network and in some cases these individuals remove themselves from the network through absence and sickness. Their absence can cause setbacks to achievement of the network's objectives. The ideal situation in a well-performing network is a spread of comparable betweenness values. Some people may indicate high betweenness but lower degree centrality, and these individuals are key figures in connecting otherwise separated groups. These boundary spanners have access to information and ideas from clusters outside their immediate groups, and are not only able to see a much bigger picture but are also able to filter information flowing from one group to another.

High closeness centrality measures indicate good access to informal information such as ideas, opinions, beliefs, and events. Individuals with strong closeness are influential in the network as they lead in forming norms and attitudes, and hence contribute greatly to building the culture of the group. The role and actions of people with strong closeness centrality should be studied in the analysis of risk in light of the objectives of the network. As opinion leaders such individuals are key to change, so their inclusion in planning, problem solving, idea generation, and change management is prudent.

Individuals who display high eigenvector values have the power to connect with many other influential people outside their immediate connections. In organizational risk terms they have the ability to form a social elite within the group, building norms and expectations that others in their group will relate to. These norms and values may, or may not, be in line with the desired culture of the overall organization network. Knowing who and how individuals with high eigenvector influence others will aid those investigating network risk or building an organization culture.

Check on Learning

Consider Figure 9.1 and Table 9.1.

1. Who in this network is high in degree? Who of these people are the most important and/or critical?

2. Who in this network has high betweenness and how much power and influence do these people have in the network?

3. Who in this network is high in closeness and how influential are these individuals in the network?

4. Who in this network is high in eigenvector and what influence do they have?

5. Who in this network is highest in all types of centrality? What risks should their manager be mindful of?

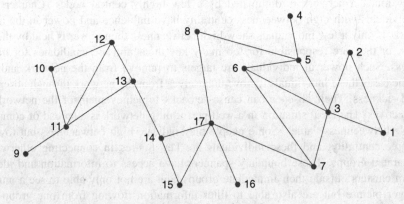

FIGURE 9.1   Undirected network graph.

TABLE 9.1   Unscaled Centrality Measures

Agent	Degree	Closeness	Betweenness	Eigenvector
1	4	0.017	0	0.452
2	4	0.017	0	0.452
3	10	0.022	58	1.000
4	2	0.016	0	0.239
5	8	0.021	38	0.767
6	6	0.021	2	0.785
7	6	0.024	72	0.756
8	4	0.023	36	0.439
9	2	0.014	0	0.045
10	4	0.014	15	0.086
11	7	0.018	62.5	0.144
12	4	0.012	6.5	0.134
13	5	0.021	96	0.198
14	6	0.025	110	0.359
15	4	0.022	0	0.312
16	2	0.020	0	0.200
17	10	0.027	144	0.643

Answer

1. Agents 3 and 17 have the highest total degree centrality with 10 followed by Agent 5 with a degree of 8. Whereas Agents 5 and 17 have links to people outside their immediate groups, Agent 3 is powerful within his/her own group, but not necessarily beyond that. Agents 5 and 17 are influential for a wider section of the network as they have links with agents in several subgroups. Agent 3 is also part of the largest number of cliques, thus adding to his/her influence.

2. Agent 17 has the highest betweenness score of 144, followed by Agent 14 with 110. These two agents are on the paths between most of the other nodes in the network and have the best opportunity to filter or change information flowing to others in the network. Both agents are targets for removal from the network to make it fragment and cease the flow of information. Agents 13 and 14 are high in betweenness and low/average in degree signifying that this person is important to the ongoing connection of their two adjacent groups. Agents 14 and 17 are funnels through which all communications travel across this network, and there is the possibility that communications can be delayed, changed, or stopped at these points in the network.

3. Agent 17 is the person who is closest to all other agents in the network, followed then by Agent 14. These opinion leaders are in prime positions to spread ideas or information across the network and need to be "onside" regarding new directions and changes in organizational activities. If these agents do not support new ideas then their influence will hamper acceptance.

4. Agent 3 has an eigenvector of 1.000, indicating this agent is very well connected to other well-connected agents. Agent 6 follows with 0.785, Agent 5 with 0.767, and Agent 4 with 0.756. With such connections these agents can easily influence group norms and behavior. The values and objectives of these individuals need to be in line with that of the organization, or there is risk that they could build competing or disparate cultures and intentions.

5. Agents 3, 5, 6, and 17 are key individuals in the network, consistently appearing high on the list in the centrality measures. Although Agent 3 is not high in betweenness, his/her influence over adjacent individuals across the network is considerably higher in centrality measures than the other people in this network. It is important to review their roles, their responsibilities, and their workloads in order to ensure there is balance in the network. People who are overloaded are single points of failure in the social network, those in boundary spanner positions filter the flows from one group to the other, opinion leaders influence the norms and values within groups, and those who are isolated or poorly connected have little idea of what is going on and are unlikely to be aligned with organizational values and goals. It is more desirable to have individuals in harmony with the organizational culture and goals in influential positions in the network to ensure effective change management, information flows, so that ideas are generated, and there is collaboration and situational awareness across the network.

People are the most valuable asset in an organization, as people have knowledge, they complete tasks, manage resources, and so forth. People are thus essential components for in-depth analysis in organizational networks together with knowledge, resources, roles, etc. for consideration in relation to risk. Take the example of removal of key people from the network—if a key person is a boundary spanner their removal could result in fragmentation of the network and groups or clusters becoming isolated from the main network. When a node is removed all the links in and out are also removed. This would result in information flows not reaching the isolated group and the communication between nodes becoming completely localized. If the person removed had access to key resources or knowledge, then the isolated cluster would no longer have access to these. The removal of links between nodes also poses a threat to achieving organizational objectives. Key links can provide bridges between clusters and changing the value or character of a link can also result in clusters becoming isolated from others in the network.

The networks in Figure 9.2, Figure 9.3, and Figure 9.4 illustrate a small social network. Bill (the circled node) in Figure 9.2 is a boundary spanner, connecting three groups of nodes. Bill is a key figure in the diffusion of ideas and information, but he could also be a choke point or may pass on a particular view of the information or even misinformation. In a disease contagion network he could be the social butterfly spreading the disease due to his links. If this network was a department in an organization, moving Bill to another department will remove him from this network. Similarly if this was a criminal network, the police could remove Bill from the network by jailing him for some illegal activity thus achieving the same result, fragmenting the network. Figure 9.3 shows the networks after

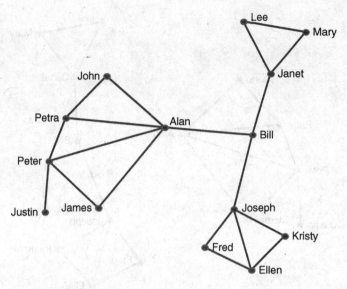

FIGURE 9.2   Social network before the removal of Bill.

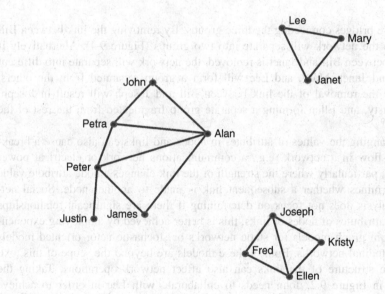

FIGURE 9.3   Fragmented social network after the removal of Bill.

Bill is removed: it has fragmented into three separate groups. In order to function more efficiently again, it will be necessary to develop new links or replace Bill with another person.

Removal of links can also fragment a network and the more important the link, the more fragmentation will result. The links between Bill and Alan, Janet and

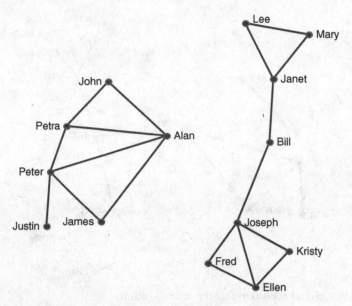

FIGURE 9.4   Removal of the link between Bill and Alan.

Joseph are bridges connecting the three groups. By removing the link between Bill and Alan the network will separate into two groups (Figure 9.4). Alternatively if the link between Bill and Janet is removed, the network will separate into different groups and Janet, Mary, and Lee will form a group separated from the others. Similarly the removal of the link between Bill and Joseph will result in Joseph, Fred, Kristy, and Ellen forming a separate group fragmented from the rest of the network.

Changing the values of attributes in nodes and links can also cause a breakdown of flow in a network (e.g., a communications network or electrical power network), particularly where the strength of the link changes a node attribute value and determines whether a subsequent link is made to another node. Social network analysis does not focus on determining if there are significant relationships between attributes of nodes or links; this is better achieved by employing exponential random graph models for static networks or stochastic actor oriented models for longitudinal networks. Both of these models are beyond the scope of this text.

The structure of the links can also affect network operations. Taking the network in Figure 9.2, John needs to collaborate with Lee in order to achieve one of his goals, but the current structure means John has to communicate with Lee via Alan, Bill, and Janet. If any of these links breaks down, then John and Lee are not able to share information. If John and Lee were to decide to make a direct connection with one another, the power of Alan, Bill, and Janet would be diminished. Alan's power in particular would be diminished if John no longer needed to link with him (Figure 9.5). Although John and Lee would more easily achieve their goals this new route may bypass certain checks and balances imposed on the previous journey through Alan, Bill, and Janet. It also means that crucial information may be lost by the interim nodes as the connection no longer takes that path.

Let us enlarge this network so we now include knowledge, resource, and task nodes associated with a few people. Sometimes, a person who is on the periphery of a social network has unique knowledge, resources, skills, etc. that make them valuable to the organization in ways that social network analysis does not detect. Assume Figure 9.5 is a network of terrorists and their task is to produce bombs (Figure 9.6). James, who has relatively low centrality, has access to the raw

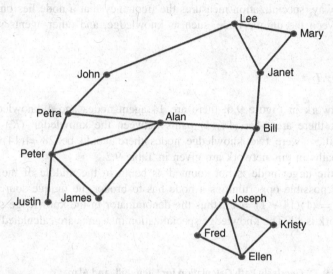

FIGURE 9.5   John and Lee make a direct link.

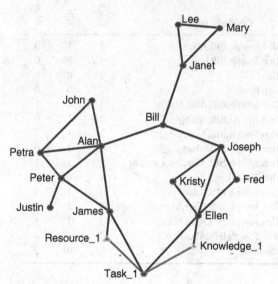

FIGURE 9.6   Linking of knowledge and resources for Task 1.

materials (Resource 1). Access to the knowledge to make the bombs is held by Ellen (Knowledge 1), who also has low relative centrality. In order to produce the bombs (Task 1) James needs to collaborate directly with Ellen to bring together the knowledge and resources to complete the task so links must be established for this to take place.

Specialization is a network measure that describes the importance of a node in terms of their unique access to knowledge, resources, tasks, etc. Specialization is calculated much like betweenness centrality. Recall that betweenness centrality measures the frequency that a node lies on potential paths between other nodes. In the same way, specialization measures the frequency that a node lies on potential paths between the unique node, such as knowledge, and other agents within the network.

Example 9.1

In the network in Figure 9.6, there are 14 agent nodes and 1 knowledge node. Therefore, there are $14 = (14)(1)$ paths between the knowledge (K) and each agent. If there were two knowledge nodes there would be $28 = (14)(2)$ paths. Shortest paths in this network are given in Table 9.2.

As the target node is not counted as being in the middle of the path, the maximum possible opportunities a node has to broker the unique source node is $K(A - 1) = (1)(14 - 1) = 13$, thus the denominator in the knowledge specialization network is 13. The knowledge specialization measures are calculated for Ellen,

TABLE 9.2 Geodesic Path Calculation for Ellen, Bill, and Alan

Path number	Source	Target	Geodesic path(s)	Ellen	Bill	Alan
1	K	Alan	{K,Ellen,Joseph,Bill,Alan}	1	1	0
2	K	Bill	{K,Ellen,Joseph,Bill}	1	0	0
3	K	Ellen	{K,Ellen}	0	0	0
4	K	Fred	{K,Ellen,Fred}	1	0	0
5	K	James	{K,Ellen,Joseph,Bill,Alan,James}	1	1	1
6	K	Janet	{K,Ellen,Joseph,Bill,Janet}	1	1	0
7	K	John	{K,Ellen,Joseph,Bill,Alan,John}	1	1	1
8	K	Joseph	{K,Ellen,Joseph,Bill,Alan}	1	1	0
9	K	Justin	{K,Ellen,Joseph,Bill,Alan,Peter,Justin}	1	1	1
10	K	Kristy	{K,Ellen,Kristy}	1	0	0
11	K	Lee	{K,Ellen,Joseph,Bill,Janet,Lee}	1	1	0
12	K	Mary	{K,Ellen,Joseph,Bill,Janet,Mary}	1	1	0
13	K	Peter	{K,Ellen,Joseph,Bill,Alan,Peter}	1	1	1
14	K	Petra	{K,Ellen,Joseph,Bill,Alan,Petra}	1	1	1
			Total Sum ⇒	13	10	5

Bill, and Alan:

$$S_{k,Ellen} = 13/13 = 1.0$$
$$S_{k,Bill} = 10/13 = 0.77$$
$$S_{k,Alan} = 5/13 = 0.38$$

Ellen brokers knowledge to the rest of the network. Although her betweenness centrality is 0 in the social network, her power comes from her unique access to knowledge. Bill and Alan do not have the same knowledge as Ellen; however, their centrality in the network allows them to broker Ellen's knowledge to others in the network. It might be the case that if all knowledge throughout the network were included in the model, that Bill and Alan would have higher relative specialization scores than Ellen. For example, if John had an equally important sole access to a different type of knowledge, similar to Ellen, then there would be two knowledge nodes. This would increase the number of geodesic paths from 14 to $(2)(14) = 28$ paths. However, Ellen would still only broker her unique knowledge. This would lower her specialization score to $13/26 = 0.5$, meaning she brokers half the knowledge in the network. Alan and Bill, however, would still have a role in brokering John's knowledge in the same manner that they brokered Ellen's knowledge. In this way, Alan and Bill's centrality in the network is observed.

Suppose we are now on the side of the peacekeepers and we wish to stop the production of the bombs. In order for Task 1 to be jeopardized we need to remove either key nodes or links. We can remove either James and/or Ellen nodes, or remove Knowledge 1 and/or Resource 1 nodes. Alternatively we can sever the links so that the key people, resources, and knowledge never get to the task. In a practical sense this could mean stopping the transport of goods or knowledge or redirecting it away from the destination of the bomb-building task. In many cases the peacekeepers will have limited resources to deploy and restrictions on their activities, so analysis is needed to ascertain the most effective and efficient means of stopping Task 1 with the resources at the peacekeepers' disposal.

In meta-networks we can study the actual connections of knowledge, resources, and roles, etc. with those that are needed in order to carry out tasks, ensuring the people with the appropriate access to resources and knowledge are allocated to specific tasks. Network analysis measures have the ability to highlight weaknesses in the allocation of knowledge, resources, tasks, and roles. Risks can be highlighted where specialist knowledge or resources are required to carry out given tasks and these are not reflected in the network connections. Imbalance and risk are evident in networks where important knowledge and resources are held by only a small number of members. Individuals with high knowledge specialization are often vital to the operations of the network and are also targets for the availability or obstruction of mission critical information or assets.

Check on Learning

An organization network consists of agents, knowledge, task, and resource nodes. Agents are assigned tasks that require specific knowledge and access to resources. List five specific threats that could be faced by this network.

Answer

Many threats could be listed, here are just a few:

1. Removal of key agents.
2. Removal of key resources.
3. Removal of key knowledge.
4. Assignment of tasks to agents who do not have the required knowledge.
5. Assignment of tasks to agents who do not have access to the required resources.
6. One or two agents having exclusive knowledge or specializing in knowledge areas.
7. One or two agents having exclusive access to resources.
8. Agents having high workloads reflected by their links to resources, knowledge, and/or tasks.
9. Bottlenecks in the network structure where boundary spanners have control over what flows in and out of subgroups.
10. Underperforming or isolated agents.

9.3   OTHER RISK MEASURES

In addition to those situations discussed earlier there are many other events that can threaten the nonachievement of objectives, including underperforming individuals, unconnected individuals, unbalanced workload, inefficient distribution of resources and knowledge, and duplications and omissions in task allotment. Network analysis offers an ability to identify vulnerabilities in an organization network's design structure based upon analyses of the node position in the network and the relationships among the people, knowledge, resources, and tasks entities. For example, Reminga and Carley (2003) and Carley et al. (2011) suggest the following as some measures of risk in organizational environments: cognitive demand, hub and authority centrality, workload, specialization and exclusiveness, situational awareness, weak points, isolates, Simmelian ties, and fragmentation.

Cognitive demand identifies agent nodes that are well situated in the network with a large number of links to other agents, tasks, and events, but lack the connectivity to knowledge and resources required to carry out their complex tasks. This requires the agent to connect to others to acquire the needed knowledge and resources to meet their requirements and fulfill their objectives. The ORA software names this measure Emergent Leader and calculates it on agent by agent matrices. Agent nodes with low cognitive demand are noted in ORA as the least integrated overall identified by the cognitive effort each agent invests to complete its tasks, connect to others, and the like. These nodes are unlikely to become formal leaders of the group.

Workload measures the extent of the knowledge and resources an agent uses in order to perform their required tasks. An agent with a high workload is involved in complex tasks and has the resources and knowledge to complete those tasks. Complex tasks are those with high demand for knowledge and resources. The workload measure is calculated in ORA from matrices of agent × knowledge, agent × resource, agent × task, knowledge × task, resource × task, and is only calculated if they all exist.

Complete exclusivity identifies those agent or organization nodes that have ties to knowledge, resources, locations, tasks, or events that other entities do not have. In contrast, specialization identifies agent or organization nodes that have ties to knowledge, resources, locations, tasks, or events that are comparatively rare in the network, that is, that few other entities have. For example, Ellen in Figure 9.6 has complete exclusivity, because her knowledge specialization is 1.0. Bill and Alan, however, do not have complete exclusivity, even though they have relatively high specialization in the network. These measures are calculated respectively using the agent × knowledge, agent × resources, agent × locations, agent × tasks, and agent × events matrices.

Isolates are nodes isolated from the rest of the network, having no links to any other node in the network. Although a node may be an isolate in the network under study, the node may be well connected in other networks that are not included in the scope of the network being analyzed.

In some networks there are nodes that are critical to the continued cohesion and operation of the network. Removal of these critical nodes will fragment the network. There are also nodes that are critical for diffusion of information across the network. Removal of these critical nodes will stop the flow of information resulting in nodes on the periphery of the network not receiving potentially critical information. Bill was a critical node in the network in Figure 9.2. It is important to be able to identify both types of critical nodes. Nodes that are critical to the continued cohesion of the network are identified in ORA and other automated social network analysis tools by determining those nodes whose removal most fragments the network. This fragmentation measure is determined by the number of nodes that can reach each other. Nodes identified to be critical to diffusion across the network are those that reach all other nodes in the network. This is calculated in ORA by measuring the distance weighted reach, which is the average distance to all other nodes in the network. Figure 9.7 illustrates a network showing the spread of an interesting image attached to an e-mail. Determination of the critical set to ensure maximum diffusion identifies nodes 8, 52, and 69 to be the most important for the spreading or dissemination of the image. These nodes are circled in Figure 9.8. This critical set is different from the set of nodes that will most splinter the network if removed. Removal of nodes 3, 8, and 65 will result in maximum fragmentation, and these three nodes are circled in Figure 9.9. Whereas the critical set for diffusion

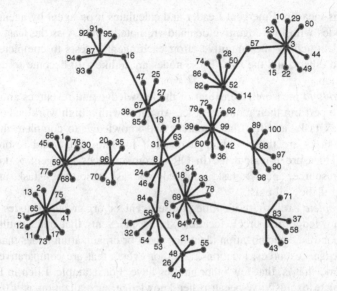

FIGURE 9.7 The network showing the spread of an interesting image by e-mail.

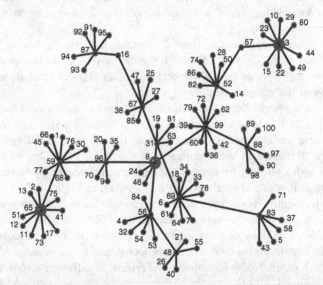

FIGURE 9.8 The three most critical nodes to the diffusion across the network, nodes 3, 8, and 65.

concentrates on the crucial nodes required to disseminate the information efficiently through the network, the fragmentation critical set concentrates on those nodes that can effectively partition and break the network apart.

Isolation of groups within the network can produce strong, insular beliefs. The lack of connection with nodes outside the group can result in "group think" where

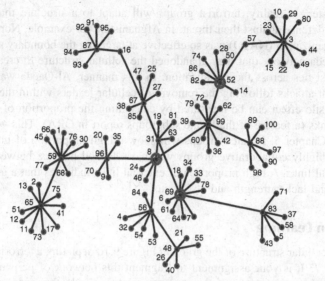

FIGURE 9.9   The three most critical nodes if removed would most fragment the network, nodes 8, 52, and 69.

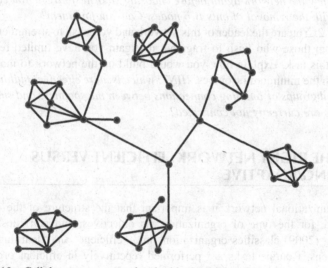

FIGURE 9.10   Cellular structure associated with clandestine networks.

the group leader becomes the "opinion leader" and members follow in order to be included as part of the group. This type of network structure is often associated with terrorist groups (Figure 9.10), where cells work in isolation, connected through opinion leaders who act as boundary spanners. This isolation can aid the generation and reinforcement of extreme or uninformed views and beliefs as the links are confined to internal communications, which strengthen the social norms and values of the cluster. In reality, terrorist groups will adapt to a structure that provides maximum defense against their threat. In Afghanistan, for example, North Atlantic Treaty Organization (NATO) was so effective at targeting the boundary spanners in the Al-Qaeda network that they abandoned the cellular structure to create thicker, more robust ties across the organization. In this manner, Al-Qaeda was still able to carry out attacks following the removal of cellular leaders within the network.

The silo effect can be measured by calculating the proportion of internal to external links (a feature of the locate subgroups report in ORA). This was demonstrated in Chapter 5. Isolated groups will show a high proportion of internal-only links and highly collaborative groups will have a good balance between internal and external links. A high proportion of external links indicates that a group is not cohesive and lacks strength and also focus.

Check on Learning

Study the cellular structure of the graph in Figure 9.10 depicting a terrorist network.

Task 1: It is your assignment to fragment this network of perpetrators. You have limited resources to get this task completed. Circle the nodes you would remove in order of their removal until the network is just a set of unconnected subgroups. *HINT: remove the boundary spanner that links the most subgroups first and then view the network again before choosing your next target. You can achieve the goal with the removal of only two nodes. Can you find them?*

Task 2: You are the leader of this network and you wish to strengthen it against attacks from those who wish to fragment it. Again you have limited resources to complete this task. Explain how you would build up the network to make it more robust with the minimum resources. *HINT: which is more effective: building bridges between subgroups or building connections between the spanners and subgroups to which they are currently not connected?*

9.4   THE RIGHT NETWORK: EFFICIENT VERSUS LEARNING/ADAPTIVE

In an organizational network it is important that the structure of the network is appropriate for the type of organization, its objectives, and activities. Daft and Armstrong (2009) classifies organizations into "efficient" and "learning/adaptive" organizations. Routine tasks are performed repetitively in efficient organizations with only the occasional need to face new challenges. An efficient hierarchical structure with centralized decision making is commonly the preferred structure of these efficient organizations. For example, organizations that manufacture electronic devices are frequently structured "efficiently" to maximize production levels. Efficient organizations prefer well-documented procedures and formal reporting. Managers will be more central in the organization, reflected by high centrality measures.

In contrast a learning organization will be dealing with new challenges frequently, requiring adaptability and problem-solving capability. The need to adapt and solve problems to succeed requires greater collaboration, sharing of knowledge and resources, and more decentralized decision making. Centrality measures for individuals in a learning organization will be higher and more evenly distributed across the network. This is due to the need for increased linkage between all network components. An efficient organization structure (with high centrality of senior managers and low centrality of employees at lower hierarchical levels) will present risk and vulnerability to a learning organization.

To illustrate the difference between efficient and learning organizations, consider a company with 400 employees and an owner. If all employees had direct access to the owner, he would quickly exceed his capacity to maintain useful links with anyone. Recall the limitations on the number of social links a person can maintain (Chapter 4). However, the link the owner chooses to drop might be the one that connects him to the next innovative idea of solution to a problem. The goal of an efficient organization is to assign managers who can review ideas and decisions and determine if they are worth the decision maker's time. Unfortunately, the manager might misinterpret the owner's goals or simply make a mistake. Thus, most organizations must strike a balance between efficiency and learning/adaptive.

Figure 9.11 illustrates a network representing an IT security applications development organization, a firm based upon a real small-medium enterprise developing specialist security software. We will call this organization InterSec Pty Ltd. (a pseudonym). This is an efficient organization, comprising 17 agents, 9 resources, 12 types of knowledge, and 14 tasks. The organization structure needs to be efficient in Daft's terms due to the repetitive nature of the work, with clear procedures and quality controls. The tasks are segmented and each task requires specialist knowledge. As is common in efficient organizations, the manager is highly central

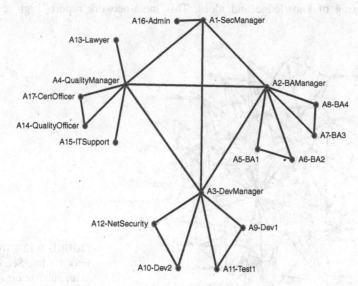

FIGURE 9.11   The efficient structure of InterSec Pty Ltd. in the agent × agent network.

to the activities, the relationships are minimal and centered upon completing tasks, and little collaboration takes place outside each worker's area of responsibility.

The hierarchical decision-making structure of InterSec Pty Ltd. is evident in Figure 9.11, with relationships only between those working together, between supervisor and worker and managerial reporting. Centralized decision making is evident, involving the more senior managers who are also those at the center of hubs; A1-SecManager, A2-BAManager, A3-DevManager, and A4-QualManager. The firm's allocation of knowledge, resources, and tasks to development projects reflected a logical, sequential, and traditional approach as reflected in Figure 9.12. In this diagram the circles represent agents, triangles represent resources, five-sided pentagons denote knowledge areas, and six-sided hexagons represent tasks.

This organization noticed that their growth was not in line with other similar organizations in their field, and they wanted to attract and produce more innovative security products. Senior management believed they had excellent staff with exceptional skills and knowledge, and wanted to encourage more collaboration and innovation in their application designs. By restructuring reporting lines and increasing collaboration InterSec formed a more dense and integrated structure, one that generated more opportunity for the spread of ideas and innovative solutions. Their new structure was more in line with Daft's learning organization. Figure 9.13 illustrates the agent by agent network for the new organizational structure and Figure 9.14 shows the new structure for the same individuals, resources, knowledge, and tasks. The increased number of links is prominent, reflecting more collaboration between individuals resulting in a greater sharing of knowledge and more distributed decision making. Notice that the managers are now no longer the only major hubs.

Comparing the two structures shows the efficient organization in Figure 9.11 has less links across the network, evidence of less interaction and integration and less sharing of knowledge and ideas. This meta-network reports high levels of

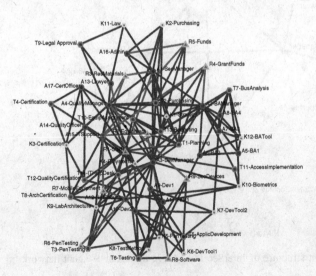

FIGURE 9.12 Total network for InterSec Pty Ltd. as an efficient organization structure.

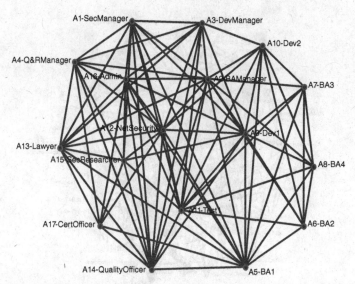

FIGURE 9.13 Learning organization structure in the agent × agent network for InterSec Pty Ltd.

exclusive knowledge, resources and tasks, and low workload, situation awareness and weak knowledge, resource and task points, particularly for the nonmanagement employees, thus increasing the risks faced by the organization. Senior managers make most of the decisions and are central to the network and risk being single points of failure. If these central nodes are removed the network is likely to fragment into unconnected subnetworks, thus seriously affecting performance levels.

The density of the network for the efficient organization structure is smaller than that for the new learning structure (Figure 9.14), which contains a greater number of hubs and a higher density overall reflecting a greater level of communication, collaboration, and knowledge sharing.

The more decentralized structure of the adaptive organization structure solves the single point of failure problem, reducing and spreading the risk. Specialization in key areas of knowledge, resources, and tasks is reduced and there is increased situation awareness and more balanced workload. There is increased flexibility in the new structure enabling regeneration of lost nodes or links due to the greater cohesion reflected by the enlarged number of links.

9.5   NETWORK THREATS AND VULNERABILITIES

As each organization will be different, there is no common list of threats, so a risk analysis should be carried out to identify possible threats to the network in question and possible changes that could minimize these threats. Four common organizational threats are presented with useful network analysis measures that may help identify and inform strategies to mitigate organizational risk.

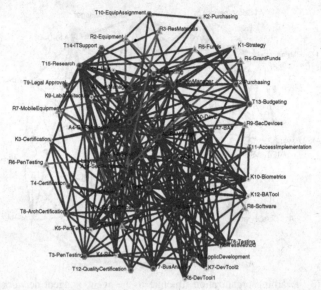

FIGURE 9.14   New total network for InterSec after changing structure to a learning organization.

9.5.1   Threat 1: Accessing Organizational Assets

As organizations become large, it is no longer possible for someone to know everyone in an organization or to know who has important resources, knowledge, skills, experience, etc. We define large as greater than 250 agents in the network. This threshold is based on the cognitive limitations a human has in maintaining social links. Agents in the organization will naturally optimize their perceived value in social connections by brokering, specialization, and experience. However, this may lead to a suboptimal solution, or individuals may attempt to gain and hold informal power in ways that benefit them, but are detrimental to the organization as a whole. Four social network measures may provide indicators of this type of threat.

High centralization will show that a few individuals dominate the network. Betweenness centralization will show power imbalance. Closeness centralization may reveal risks for communication across the organization. High centralization must be interpreted carefully, however. Highly central individuals could be serving a role to connect individuals with organizational assets in a positive way or they could be acquiring high informal power in a manner that is not consistent with the organization's direction. Depending upon the organization's size and purpose, the organization may need these individuals. In this case, there should be multiple actors filling this role so that the removal of a key node does not seriously impact the organization's ability to function.

An agent's specialization measure can also show how a few individuals can dominate the network in a slightly different way. Individuals high in specialization might have skill sets that are highly expensive to replicate in terms of education or experience. In this situation, it may not be practical to hire additional individuals to reduce the organizational risk. However, if it is an issue of organizational education/experience, training and professional development programs can help diversify knowledge, skill, and resource awareness across the organization.

High fragmentation will show that the network is vulnerable to having critical organizational assets cut off from the rest of the organization. This might involve outsourcing something, where the organization has an idle capability. It might result in lost sales due to ignorance of the sales staff. The organization might even be paralyzed as they are forced to rebuild connections within the organization to function. Fragmentation is rarely beneficial in an organization and usually requires concerted effort to thicken the network.

A large diameter is another measure that can show significant distance from one end of the organization to the other. Managers might be able to develop relationships between distant ends of an organization to reduce the diameter and thus increase the ability of agents to access organizational resources. As a rule of thumb, a diameter less than or equal to three has ready access to all organizational resources. This is based on research on cognitive social structures, which shows that an individual knows their friends (one degree), has a pretty good idea of their friend's friends (two degrees), and might know some of their friend's friend's friends (three degrees). Beyond three degrees of separation, however, the likelihood of awareness approaches zero. Thus, a diameter no larger than three provides reasonable awareness of other agents in the organization.

9.5.2 Threat 2: Cognitive Overload

Efficient organizations have hierarchies that are meant to enable high degrees of supervision and protect the time of decision makers by delegating authority to lower-level managers. These activities can sometimes result in certain managers being overloaded/overworked, while other managers are relatively unburdened. This situation is not an optimal use of employees, and in some cases can lead to organization risk if the overloaded manager resigns, is sick, or is otherwise removed from the network. Cognitive overload is detected with high network centralization and high cognitive demand.

9.5.3 Threat 3: Silo Effect

Learning adaptive organizations are meant to facilitate collaboration and interaction across the group. A silo effect occurs where a portion of the organization is isolated from the larger group producing a silo. Silos inhibit collaboration and prevent learning organizations from functioning as intended. The silo index, presented in Chapter 5, identifies subgroups of the organization that tend to be isolated and insular. Reducing these silos requires deliberate thickening of the network, specifically oriented at the agents in the silo subgroup. The fragmentation measure may also provide indications of silos.

9.5.4   Threat 4: Unwelcomed Organizational Change

Organizational culture is the set of values and beliefs held by members of the organization. Typically, this consists of things such as the value placed on education versus experience; importance of the worker versus the manager; how to deal with customers, or the value an employee provides evaluated by time at work versus some other performance measure. Informal leaders in the network occupy positions of opinion leadership and are able to alter the organizational culture and implement change. In previous chapters, we alluded to the importance of identifying these individuals and seeking their buy-in for important decisions. These individuals also have the ability to implement unwelcome organizational change. Informal leaders are those high in one or more of the node-level centrality measures. These individuals should be identified and managed to ensure the desired organizational culture is maintained.

9.6   THICKENING A NETWORK

Managers may implement strategies to *thicken a network* to create redundant connections and increase collaboration. This is very desirable for knowledge-intensive organizations and learning/adaptive organizations. It may not be good to thicken the network in an efficient organization, because each connection requires time and resources to maintain the links and can distract from the task an individual is intended to perform. Strategies for thickening a network are based in the theories of social link formation presented in Chapter 4. It is important to keep in mind that relationships and affinity cannot be forced. Therefore, a manager must simply create opportunities for link formation and interaction. This will come at a cost of less time spent on an actual billable task; however, the thickened network provides increased value to the organization as a whole. This is often called *social capital*.

Proximity is the simplest approach for thickening a network. At Apple, there was a single bathroom that employees had to use. Many Apple employees have a "bathroom story" where they met someone from a diverse part of the organization. During US military operations following the US–Iraq war, the Northern Division established a coffee bar, where soldiers in the division headquarters could get free coffee. This had an observed effect of significant thickening of the social network and increased collaboration across the staff.

Proximity can be physical. In this sense, creating an open office plan or assigning people to desks/cubicles near each other can increase their opportunity for interaction.

Proximity can be organizational. People can be assigned to teams. The team leader will naturally facilitate collaboration through meeting, working groups, etc. These teams can be formally assigned as in a divisional organizational structure, or they can be organized in ad hoc working groups as in a functional organizational structure.

The most effective structure for increasing collaboration is to matrix the physical and organizational structures. Do not assign people to working locations near other people in their same organizational structure. If people are assigned to product lines, for example, do not sit them together. Instead, have employees from different product lines sit near each other. This will increase the effort that a product manager will expend to manage his personnel who are dispersed across the office; however, it will provide greater opportunity for those employees to meet people working on different product lines, gain greater situational awareness, exchange ideas, and expand their individual social capital.

Homophily can be used to thicken the network by making certain traits known to others. Some organizations use a Web site to highlight individual employee profiles. Military uniforms provide a wealth of information about service members, what theaters of operation they have deployed to, special skills such as airborne or combat diver. Social forums are sometimes offered such as a weekly bible study or exercise group. These strategies all take advantage of homophily to increase the social capital of individuals and thicken the social network.

Transitivity and balance can also provide mechanisms for increased collaboration. Ad hoc project teams and short-term projects may provide opportunities to put people from diverse parts of the organization together. These individuals may introduce members of the ad hoc team to others in their long-term section based on transitivity. If there are issues of affinity and dislike, carefully constructed team assignment based on balance theory may help alleviate organizational conflict.

There are endless methods to increase the volume of connections in an organization. This is typically more advantageous for learning/adaptive organizations than for efficient organizations. All strategies will consume resources, so the value realized in a thicker network must justify the expense. This section highlights some easily implemented strategies for thickening a network. The reader is encouraged to review Chapter 4 and reason through additional strategies that may increase collaboration in their organization.

9.7   THINNING A NETWORK

Thinning a network is essentially the inverse problem of thickening a network and it is also based in the theories of social link formation presented in Chapter 4. Efficient organizations may wish to thin a network to increase time spent on task performance and reduce social interaction. In criminal and military applications, thinning a network might be conducted to disrupt or degrade the performance of a network.

Proximity can be leveraged to thin a network. Increased physical barriers will reduce opportunities for interaction and relationships will atrophy. Creating physical barriers could be as simple as closing off a doorway or removing a common break room. Organizational barriers could be created through increased management reporting layers. Communication outlets could be restricted through blocked Web sites, reduced phone access, restricted phone numbers that can be dialed on particular phones.

Another approach for thinning a network is to remove highly central nodes or disrupt central links. Identifying boundary spanners who are low in degree and high in betweenness can be particularly effective. If highly central nodes cannot be removed, they can be more closely managed and influenced as opinion leaders and informal leaders. Informal leaders that cannot be managed should be removed through relocation, promotion, or lay-off. Beware that the perceived treatment of informal leaders will have a definite impact on the organizational culture. Informal leaders may also continue to wield influence in the organization after they have officially left the group.

Enhancing team identity or having subgroup competition can increase the subgroup identity over the larger organizational identity. A strong team identity can create perceptions of within team homophily and heterophily with those outside the subgroup. This can increase a silo effect within teams. As with thickening a network, there are many ways to thin a network. Strategies are based on theories for social link formation presented in Chapter 4. The reader is encouraged to reason through other strategies that may thin a social network.

9.8   PROCESS OF ORGANIZATIONAL RISK ANALYSIS

Network analysis can assist in the identification of potential threats and is valuable to include in the early stages of risk analysis and management. Social network analysis also has a role in the development of mitigation strategies, permitting simulation of situations while investigating the impact of actions that change the network structure and dynamics. The following activities provide a useful approach for the application of social network analysis to organizational risk. Although this set of activities does not form a rigorous methodology, it can be a valuable guide in studying an organizational situation of interest using network analysis. Figure 9.15 has been drawn from these activities and illustrates the areas that need to be considered in analyzing risk in organizations using network analysis.

1. *Aim and Perspective.* Determine your aim in studying this network. What is your role and from whose perspective are you studying the network? Who else could be interested in this network, what stakeholding do they have and how would they perceive the network? This holistic analysis will provide a much bigger picture of the network under study in its environment increasing situational awareness and enable greater understanding of internal and external links as well as potential risks that may emanate from both within and outside the network. By looking critically at a network an investigator can see how information about a network can be used and misused depending upon one's alliance or stakeholding. This activity influences the other activities in the risk identification process as a given perspective will influence the interpretation of the findings in each stage.

2. *Network Questions and Data.* Who created this network and who is the owner of this network? What type of organization is it, and what are its objectives? What are the questions you seek to answer in this analysis? What data collection methods were used? Who defined the node types, links, and link values?

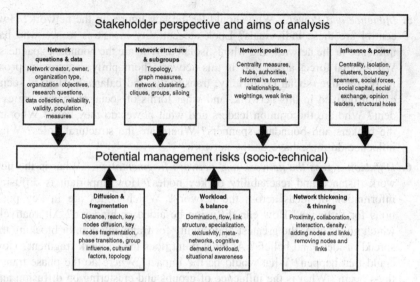

FIGURE 9.15 Activities in identifying risks in organizations.

How reliable and representative are the nodes, links, and values to the real network? Networks can be created from a specific perspective, indicating potential subjectivity in the building of the network structure, and definition of links and data used to weigh attributes and link values. The creator of the network instance needs to be known to enable the investigator to ascertain reliability and trustworthiness of contributors to the network study.

3. *Structure of the Network and Subgroups.* Analyze the structure (topology) of the network. Is the network structure suitable for this organization's operations and goals? Analyze the groups and clusters. Does this structure fit with the topology required to achieve the objectives? Who has power in the groups? How insular are the groups? If the aim is to diffuse information how well does the structure aid this? If the aim is to segment and control, how well does the structure achieve this? An understanding of the structure of the network, its groupings, breadth and density of connections is necessary to identify risks and reduce their potential occurrence.

4. *Network Position.* Look at the centrality measures for all nodes (degree, betweenness, closeness, eigenvector). Are there hubs and authorities in the network? Is there a formal and informal social network? Who reports to whom and who do people go to for advice? Who likes who, who trusts who? Compare the informal and formal networks. Although official power is reflected in a formal network, the actual work gets down through the informal networks in organizations. It is necessary to know who holds the real power in organizational networks. What do the relationships actually tell us? What effect do the weightings of these relationships have on the network? What are the weak relationships and how can these provide strength and opportunities in the network?

5. *Influence and Power.* Who should be the key figures in the network? How central are these individuals? Look at centrality measures to see who has influence in the network and who is isolated. Who are the boundary spanners? What social forces are at play in this network (homophily, reciprocity, proximity, prestige, social conformity, transitivity, and balance)? How is social capital derived in this network, and what forms of social exchange are evident? Who are the opinion leaders and what power do they have? Who are the brokers and boundary spanners? Where are the structural holes? What influence do these have on the network?

6. *Diffusion across the network and network fragmentation.* What is the network distance and reachability of key nodes? How important is diffusing information or ideas across the network? Which nodes are in key positions for diffusion? How can diffusion be aided or hindered? Alternatively which nodes offer the greatest opportunity for fragmentation, or blocking the spread of ideas and beliefs? At what point does the network fragment? How could that happen? What would be the impact? Where do the phase transitions occur? What is the influence of groups and clustering on diffusion and fragmentation? What cultural factors need to be considered for diffusion or fragmentation? How does the topology of the network aid or hinder diffusion and fragmentation?

7. *Workload and Balance.* Investigate Potential Management Risks: Is the workload balanced? Who is overworked and who is underworked? Who is well informed and who is isolated? Who dominates in the network? How do the multiple layers of the network integrate to form the meta-network? Investigate the cognitive demand on individuals. Look at the knowledge, resources, and tasks. Who controls these and how effectively are they allocated? Who specializes in what and who has exclusivity over what? Study the level of social connection and situation awareness.

8. *Network Thickening and Thinning.* Does the network need to be thickened to make it more robust? Can the density be increased by the addition of random nodes or links, if not, what form do these additions need to take? Can the network be thickened in the right places by proximity related measures? How can collaboration and interaction between nodes be increased? How can brokers be effectively situated between clusters to thicken the network? Does the network need to be thinned and how could that be best achieved? Which nodes need to be removed for most effective impact? Which links need to be removed or changed for best effect? Perform "what if" scenarios to simulate changes in the network structure and dynamics and compare these networks over different timeframes. How do these changes contribute to meeting the organization's objectives?

9. *Look back.* Go back to the first step—have you satisfied your aims in analyzing the network?

9.9   SUMMARY OF MAIN POINTS

Risk is the effect of uncertainty on objectives. The effect of a risk occurring can be either positive or negative. Organization networks have a purpose, and risk in networks relates to the network's ability to achieve its objectives. Factors influencing network risk can arise from both internal and external sources.

Threats to social networks take the form of removal, addition, or modifications of nodes and/or links. Nodes and links are the components of networks and threats to networks emanate from changes to these components.

Measures used to investigate social network risk in organizations are the same measures used to study social network structure and activities in organizations. Centrality measures can indicate risks as well as network strengths and form the basis of other indicators that are used in network risk analysis. Position in the network and structure of groups can affect the diffusion of information and ideas as well as the network's susceptibility to fragmentation.

Potential risks in organizational meta-networks can be identified by studying the distribution of knowledge, resources, roles, and tasks. Specialization and exclusivity indicates that knowledge, resources, and roles, etc. are held by only a few, making these key people targets for removal from the network. Imbalance in the distribution of workload, weak levels of situation awareness and isolation of key nodes are risks to the network achieving its objectives.

Efficient organizational structures are commonplace in manufacturing and other hierarchical networks. These organizations have a clear reporting structure and links between individuals are predominantly with those who are members of the same groups or departments. Collaboration is not an objective, efficiency of processes is the main aim and relationships are built around the achievement of goals associated with production levels.

A learning organization on the other hand illustrates a network that is much denser than an efficient organization, as collaboration is needed in order to achieve the objectives to integrate knowledge and resources with tasks to get a job done and generate new ideas.

Thickening the network can be achieved by encouraging increased interaction with others, particularly those in different groups and developing redundant connections to increase collaboration and supporting the exchange of ideas. This can be done physically by assigning people to teams, providing an open workspace, a meeting place, and positioning individuals to maximize exchange and collaboration.

Thinning the network can be achieved by increasing physical barriers between people, blocking information flows or removing highly central nodes or dislocating central links. Individuals and resources can be moved, reassigned, disabled, or removed altogether.

CHAPTER 9   LAB EXERCISE

Organizational Risk Laboratory

Gamesmith Pty Ltd. is partnering with Raveda Pty Ltd. for a joint venture. Gamesmith is a computer game development company, who specialize in security games for corporate training. They have several "games" they use in training focused on offensive and defensive cyber warfare, penetration testing, and network security. Raveda are a consulting company specializing in competitive intelligence and market force advice. The two companies jointly tendered for a project with the government and have been successful.

The project requires the collection and analysis of intelligence through the delivery of security training via security games. The government wishes all employees within several of its key departments and business-to-business partners to undertake professional development in network security to enable intelligence data to be gathered and analyzed on their patterns of attack and defense, their strategies in cyber warfare, and use of information. Basically they are gathering data on the behavior of their employees and business partners. These data will then be fed into the national defense intelligence system to identify potential operatives for national security agencies as well as potential terrorists and insiders. The new project is called *HI3*.

Gamesmith and Raveda have combined their employees for this project. This meta-network has node types of Agent, Task, and Knowledge. Agents 1–12 are from Gamesmith and Agents 13–20 are from Raveda. There are two (Agent × Agent) networks, the formal network and the informal network for the merged project. The formal structure for the joint project looks like this:

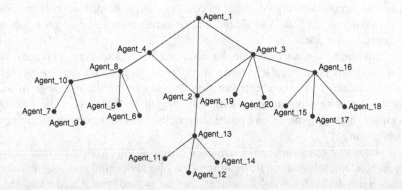

Agent_1 is the overall boss and his supervisory team consists of Agents 2, 3, and 4.

Enter the data from Table B.15, Table B.16, Table B.17, Table B.18 and Table B.19 in Appendix B into ORA and generate some reports summarizing analysis of this meta-network. The Management, All Measures, Key Entities, and Critical

Set reports will give you information you may find useful. Other reports can also provide good information in your analyses of this organization.

You are required to analyze the network and identify concerns (threats) for the executive manager who is not shown in the network, to whom Agent_1 reports. The Executive Manager has employed you to complete this report, which will be submitted directly to him. To complete this task you need to visualize and analyze the networks through both the visualizer and reports. This is a meta-network so you will need to use meta-network measures in the reports.

Some questions that may assist you in your analysis are

1. What are the threats? What are the sources of those threats?
2. How likely is the occurrence of the threat (i.e., how capable is the threat source to carry out the threat)?
3. What would be the impact if the threat was to occur?
4. How can we minimize the threat occurrence? How can we minimize the impact?
5. What changes do you recommend in order to make the people in this network more collaborative? An example of a possible recommendation would be to thicken the network to increase the diffusion of information.

Some questions you might ask in your analysis could be

- Who are the key people in the network and what may happen if they were removed from the network (sickness, accident, won lotteries, etc.)?
- Who has specialized or exclusive knowledge and what may happen if that agent or knowledge were removed?
- Who has specialized or exclusive tasks assigned and what may happen if that agent or task were removed?
- Are there any discrepancies in the allocation of tasks to agents based upon knowledge?
- What potential threats could arise from the informal network structure (e.g., power, cliques, boundary spanners, subgroups, etc.)?

EXERCISES

9.1. Choose an organization you know well—for example, the company or department where you work or a sports group to which you belong. Think about the structure of this group as a network and identify who has knowledge, who undertakes tasks, how the people interact or collaborate. How could you increase collaboration or the exchange of ideas in this network?

9.2. Identify the risks that this group faces with regard to its structure, the key people, and the interactions between people in this group. Are there any people with exclusive knowledge or exclusive control over resources? How could you reduce the risk in this network?

9.3. Think of 10 ways you could thicken a network to encourage innovation and new ideas.

9.4. Think of 10 ways you could thin a network to increase control and efficiency.

REFERENCES

Carley, K., Reminga, J., Storrick, J., and Columbus, D. (2011). Ora user's guide. Technical Report CMU-ISR-11-107, CASOS, CMU.

Daft, R. L. and Armstrong, A. (2009). *Organization theory and design*. Nelson Education.

Reminga, J. and Carley, K. (2003). Measures in ORA. CASOS, Carnegie Mellon University.

MATRIX ALGEBRA PRIMER

Definitions

A *matrix* is a collection of numbers ordered by rows and columns. It is customary to enclose the elements of a matrix in parentheses, brackets, or braces. For example, the following is a matrix:

$$\mathbf{A} = \begin{bmatrix} 3 & 8 & 2 \\ 5 & 0 & -1 \end{bmatrix}$$

This matrix has two rows and three columns, so it is referred to as a 2×3 matrix. The elements of a matrix are numbered in the following way:

$$\mathbf{A} = \begin{bmatrix} a_{11} & a_{12} & a_{13} \\ a_{21} & a_{22} & a_{23} \end{bmatrix}$$

That is, the first subscript in a matrix element refers to the row that the element resides, and the second subscript refers to the column. It is important to remember this convention when matrix algebra is performed.

A *vector* is a special type of matrix that has only one row (called a *row vector*) or one column (called a *column vector*). Below, **B** is a column vector whereas **C** is a row vector.

$$\mathbf{B} = \begin{bmatrix} a_{11} \\ a_{21} \\ a_{31} \end{bmatrix}, \quad \mathbf{C} = \begin{bmatrix} c_{11} & c_{12} & c_{13} \end{bmatrix}$$

A *scalar* is a matrix with only one row and one column. It is customary to denote scalars by italicized, lower case letters (e.g., *a*), to denote vectors by bold, lower case letters (e.g., **a**), and to denote matrices with more than one row and one column by bold, upper case letters (e.g., **A**).

A *square matrix* has as many rows as it has columns. Matrix **A** is square but matrix **B** is not square:

$$\mathbf{A} = \begin{bmatrix} 1 & 2 \\ 5 & 3 \end{bmatrix}, \quad \mathbf{B} = \begin{bmatrix} 7 & 2 \\ 3 & 6 \\ 5 & 0 \end{bmatrix}$$

Social Network Analysis: with Applications, First Edition.
Ian A. McCulloh, Helen L. Armstrong, and Anthony N. Johnson.
© 2013 John Wiley & Sons, Inc. Published 2013 by John Wiley & Sons, Inc.

A *symmetric matrix* is a square matrix in which $x_{ij} = x_{ji}$ for all i and j. Matrix **A** is symmetric; matrix **B** is not symmetric.

$$\mathbf{A} = \begin{bmatrix} 0 & 5 & 2 \\ 5 & 8 & 3 \\ 2 & 3 & 7 \end{bmatrix}, \quad \mathbf{B} = \begin{bmatrix} 0 & 6 & 2 \\ 5 & 8 & 3 \\ 2 & 6 & 7 \end{bmatrix}$$

A *diagonal matrix* is a symmetric matrix where all the off diagonal elements are 0. Matrix **A** is diagonal.

$$\mathbf{A} = \begin{bmatrix} 1 & 0 & 0 \\ 0 & 8 & 0 \\ 0 & 0 & 7 \end{bmatrix}$$

An *identity matrix* is a diagonal matrix with 1s and only 1s on the diagonal. The identity matrix is almost always denoted as **I**.

$$\mathbf{I} = \begin{bmatrix} 1 & 0 & 0 \\ 0 & 1 & 0 \\ 0 & 0 & 1 \end{bmatrix}$$

Matrix Addition and Subtraction

To add two matrices, they both must have the same number of rows and they both must have the same number of columns. The elements of the two matrices are simply added together, element by element, to produce the results. That is, for $\mathbf{R} = \mathbf{A} + \mathbf{B}$, then

$$r_{ij} = a_{ij} + b_{ij}$$

for all i and j. Thus,

$$\begin{bmatrix} 9 & 5 & 1 \\ -4 & 7 & 6 \end{bmatrix} = \begin{bmatrix} 1 & 9 & -2 \\ 3 & 6 & 0 \end{bmatrix} + \begin{bmatrix} 8 & -4 & 3 \\ -7 & 1 & 6 \end{bmatrix}$$

Matrix subtraction works in the same way, except that elements are subtracted instead of added.

Matrix Multiplication

There are several rules for matrix multiplication. The first concerns the multiplication between a matrix and a scalar. Here, each element in the product matrix is simply the scalar multiplied by the element in the matrix. That is, for $\mathbf{R} = a\mathbf{B}$, then

$$r_{ij} = a(b_{ij})$$

for all i and j. Thus,

$$8 \begin{bmatrix} 16 & 48 \\ 24 & 56 \end{bmatrix} = \begin{bmatrix} 2 & 6 \\ 3 & 7 \end{bmatrix}$$

Matrix multiplication involving a scalar is commutative. That is, $a\mathbf{B} = \mathbf{B}a$.

The next rule involves the multiplication of a row vector by a column vector. To perform this, the row vector must have as many columns as the column vector has rows. For example,

$$\begin{bmatrix} 9 & 5 & 1 \end{bmatrix} \begin{bmatrix} 1 \\ 3 \\ 5 \end{bmatrix}$$

is legal. However,

$$\begin{bmatrix} 9 & 5 & 1 \end{bmatrix} \begin{bmatrix} 1 \\ 3 \\ 5 \\ 4 \end{bmatrix}$$

is not legal because the row vector has three columns, whereas the column vector has four rows. The product of a row vector multiplied by a column vector will be a scalar. This scalar is simply the sum of the first row vector element multiplied by the first column vector element plus the second row vector element multiplied by the second column vector element plus the product of the third elements, etc. In algebra, if $r = \mathbf{ab}$, then

$$r = \sum_{i=1}^{n} a_i b_i$$

Thus,

$$\begin{bmatrix} 9 & 5 & 1 \end{bmatrix} \begin{bmatrix} 1 \\ 3 \\ 5 \end{bmatrix} = 9(1) + 5(3) + 1(5) = 9 + 15 + 5 = 29$$

All other types of matrix multiplication involve the multiplication of a row vector and a column vector. Specifically, in the expression $\mathbf{R} = \mathbf{AB}$,

$$r_{ij} = \mathbf{a}_{i\cdot}\mathbf{b}_{\cdot j}$$

where $\mathbf{a}_{i\cdot}$ is the ith row vector in matrix $\mathbf{A}$ and $\mathbf{b}_{\cdot j}$ is the jth column vector in matrix $\mathbf{B}$. Thus, if

$$\mathbf{A} = \begin{bmatrix} 2 & 8 & 1 \\ 3 & 6 & 4 \end{bmatrix}, \quad \text{and} \quad \mathbf{B} = \begin{bmatrix} 1 & 7 \\ 9 & -2 \\ 6 & 3 \end{bmatrix}$$

then

$$r_{11} = a_{1\cdot}b_{\cdot 1} = \begin{bmatrix} 2 & 8 & 1 \end{bmatrix} \begin{bmatrix} 1 \\ 9 \\ 6 \end{bmatrix} = 2(1) + 8(9) + 1(6) = 80$$

and

$$r_{12} = a_{1\cdot}b_{\cdot 2} = \begin{bmatrix} 2 & 8 & 1 \end{bmatrix} \begin{bmatrix} 7 \\ -2 \\ 3 \end{bmatrix} = 2(7) + 8(-2) + 1(3) = 1$$

and

$$r_{21} = a_{2}.b_{.1} = \begin{bmatrix} 3 & 6 & 4 \end{bmatrix} \begin{bmatrix} 1 \\ 9 \\ 6 \end{bmatrix} = 3(1) + 6(9) + 4(6) = 81$$

and

$$r_{22} = a_{2}.b_{.2} = \begin{bmatrix} 3 & 6 & 4 \end{bmatrix} \begin{bmatrix} 7 \\ -2 \\ 3 \end{bmatrix} = 3(7) + 6(-2) + 4(3) = 21$$

Hence,

$$\begin{bmatrix} 2 & 8 & 1 \\ 3 & 6 & 4 \end{bmatrix} \begin{bmatrix} 1 & 7 \\ 9 & -2 \\ 6 & 3 \end{bmatrix} = \begin{bmatrix} 80 & 1 \\ 81 & 21 \end{bmatrix}$$

For matrix multiplication to be legal, the first matrix must have as many rows as the second matrix has columns. This, of course, is the requirement for multiplying a row vector by a column vector. The resulting matrix will have as many rows as the first matrix and as many columns as the second matrix. Because **A** has two rows and three columns, whereas **B** has three rows and two columns, the matrix multiplication may legally proceed and the resulting matrix will have two rows and two columns.

Because of these requirements, matrix multiplication is usually not commutative. That is, usually **AB** ≠ **BA**. And even if **AB** is a legal operation, there is no guarantee that **BA** will also be legal. For these reasons, the terms *premultiply* and *postmultiply* are often encountered in matrix algebra, whereas they are seldom encountered in scalar algebra.

One special case to be aware of is when a column vector is postmultiplied by a row vector. Consider

$$\begin{bmatrix} -3 \\ 4 \\ 7 \end{bmatrix} \begin{bmatrix} 8 & 2 \end{bmatrix} ?$$

In this case, one simply follows the rules given above for the multiplication of two matrices. Note that the first matrix has one column and the second matrix has one row, so the matrix multiplication is legal. The resulting matrix will have as many rows as the first matrix (three) and as many columns as the second matrix (two). Hence, the result is

$$\begin{bmatrix} -3 \\ 4 \\ 7 \end{bmatrix} \begin{bmatrix} 8 & 2 \end{bmatrix} = \begin{bmatrix} -24 & -6 \\ 32 & 8 \\ 56 & 14 \end{bmatrix}$$

Matrix Transpose

The transpose of a matrix is denoted by a prime ($\mathbf{A}'$) or a superscript t or T ($\mathbf{A}^{t}$ or $\mathbf{A}^{T}$). The first row of a matrix becomes the first column of the transpose matrix, the second row of the matrix becomes the second column of the transpose, etc. Thus,

$$\mathbf{A} = \begin{bmatrix} 2 & 8 & -1 \\ 3 & 6 & 4 \end{bmatrix}, \quad \text{and} \quad \mathbf{A}^T = \begin{bmatrix} 2 & 3 \\ 8 & 6 \\ -1 & 4 \end{bmatrix}$$

The transpose of a row vector will be a column vector, and the transpose of a column vector will be a row vector. The transpose of a symmetric matrix is simply the original matrix.

Matrix Inverse

In scalar algebra, the inverse of a number is that number, which, when multiplied by the original number, gives a product of 1. Hence, the inverse of x is simple $1/x$ or, in slightly different notation, x^{-1}. In matrix algebra, the inverse of a matrix is that matrix, which, when multiplied by the original matrix, gives an identity matrix. The inverse of a matrix is denoted by the superscript -1. Hence,

$$\mathbf{A}\mathbf{A}^{-1} = \mathbf{A}^{-1}\mathbf{A} = \mathbf{I}$$

A matrix must be square to have an inverse, but not all square matrices have an inverse. In some cases, the inverse does not exist.

Matrix Algebra Exercises

1. $\begin{bmatrix} 0 & 2 \\ -2 & -5 \end{bmatrix} \begin{bmatrix} 6 & -6 \\ 3 & 0 \end{bmatrix}$

2. $\begin{bmatrix} 6 \\ -3 \end{bmatrix} \begin{bmatrix} -5 & 4 \end{bmatrix}$

3. $\begin{bmatrix} -5 & -5 \\ -1 & 2 \end{bmatrix} \begin{bmatrix} -2 & -3 \\ 3 & 5 \end{bmatrix}$

4. $\begin{bmatrix} -3 & 5 \\ -2 & 1 \end{bmatrix} \begin{bmatrix} 6 & -2 \\ 1 & -5 \end{bmatrix}$

5. $\begin{bmatrix} 0 & 5 \\ -3 & 1 \\ -5 & 1 \end{bmatrix} \begin{bmatrix} -4 & 4 \\ -2 & -4 \end{bmatrix}$

6. $\begin{bmatrix} 5 & 3 & 5 \\ 1 & 5 & 0 \end{bmatrix} \begin{bmatrix} -4 & 2 \\ -3 & 4 \\ 3 & -5 \end{bmatrix}$

7. $\begin{bmatrix} -5 \\ 6 \\ 0 \end{bmatrix} \begin{bmatrix} 3 & -1 \end{bmatrix}$

8. $\begin{bmatrix} 3 & 2 & 5 \\ 2 & 3 & 1 \end{bmatrix} \begin{bmatrix} 4 & 5 & -5 \\ 5 & -1 & 6 \end{bmatrix}$

9. $\begin{bmatrix} 3 & -1 \\ -3 & 6 \\ -6 & -6 \end{bmatrix} \begin{bmatrix} -1 & -6 \\ 5 & 4 \end{bmatrix}$

10. $\begin{bmatrix} 5 & 4 \\ 2 & -1 \end{bmatrix} \begin{bmatrix} -4 \\ 3 \end{bmatrix}$

11. $\begin{bmatrix} -1 & 1 & -1 \\ 5 & 2 & -5 \\ 6 & -5 & 1 \\ -5 & 6 & 0 \end{bmatrix} \begin{bmatrix} 6 & 5 \\ 5 & -6 \\ 6 & 0 \end{bmatrix}$

12. $\begin{bmatrix} -2 & -6 \\ -4 & 3 \\ 5 & 0 \\ 4 & -6 \end{bmatrix} \begin{bmatrix} 2 & -2 & 2 \\ -2 & 0 & -3 \end{bmatrix}$

13. $\begin{bmatrix} 2 & -5v \end{bmatrix} \begin{bmatrix} -5u & -v \\ 0 & 6 \end{bmatrix}$

14. $\begin{bmatrix} -4 & -y \\ -2x & -4 \end{bmatrix} \begin{bmatrix} -4x & 0 \\ 2y & -5 \end{bmatrix}$

Critical Thinking Question 1 Write an example of a matrix multiplication that is undefined.

Critical Thinking Question 2 In the expression AB, if A is a 3×5 matrix, then what could be the dimensions of B?

APPENDIX B

TABLES OF DATA
AND NETWORKS

TABLE B.1   Lab Exercise 1 — Agent Node Class

Agent ID No.	Title
Agent 1	Alice
Agent 2	Bert
Agent 3	Chris
Agent 4	David
Agent 5	Eileen
Agent 6	Nan
Agent 7	Cliff
Agent 8	John
Agent 9	Peter
Agent 10	Sharon

TABLE B.2   Lab Exercise 1 — Agent × Agent Adjacency Matrix

ID	A1	A2	A3	A4	A5	A6	A7	A8	A9	A10
Agent 1	0	1	1	1	0	0	1	0	1	0
Agent 2	1	0	1	1	1	0	1	0	1	0
Agent 3	1	1	0	1	1	0	0	0	0	0
Agent 4	1	1	1	0	0	0	0	0	0	0
Agent 5	0	1	1	0	0	1	0	1	0	0
Agent 6	0	0	0	0	1	0	0	1	1	1
Agent 7	1	1	0	0	0	0	0	0	1	0
Agent 8	0	0	0	0	1	1	0	0	0	0
Agent 9	1	1	0	0	0	1	1	0	0	1
Agent 10	0	0	0	0	0	1	0	0	1	0

Social Network Analysis: with Applications, First Edition.
Ian A. McCulloh, Helen L. Armstrong, and Anthony N. Johnson.
© 2013 John Wiley & Sons, Inc. Published 2013 by John Wiley & Sons, Inc.

TABLE B.3   Health Organization — Agent Node Class

Agent ID No.	Title
A1	Jan
A2	Trevor
A3	Verna
A4	Joy
A5	Colin
A6	Shar
A7	Cher
A8	Abby
A9	Kaye
A10	Lynn
A11	May
A12	Janet
A13	Laene
A14	Melanie
A15	Amy
A16	Jani
A17	Bonny
A18	Marie
A19	Helen
A20	Kerry
A21	Soshi
A22	Ravine
A23	Grace
A24	Shiswe
A25	Tess
A26	Nikkie
A27	Huang
A28	Melly
A29	Yang
A30	Rassy
A31	Beth
A32	Angela
A33	Kira
A34	Pedram
A35	Jessica
A36	Annie
A37	Mary
A38	Rana
A39	Jayne
A40	Fran
A41	Millie
A42	Carol
A43	Jean

TABLE B.4   Health Organization — Agent × Agent Admin Advice

ID	A1	A2	A3	A4	A5	A6	A7	A8	A9	A10	A11	A12	A13	A14	A15	A16
A1	0	2	0	0	4	4	3	0	4	0	0	0	0	0	0	0
A2	2	0	1	2	2	2	1	2	2	0	0	0	0	0	0	0
A3	0	0	0	2	2	2	0	2	2	0	0	0	0	0	0	0
A4	0	0	0	0	0	3	0	0	0	0	0	0	0	0	0	0
A5	0	2	2	2	0	2	2	2	2	0	0	0	0	0	0	0
A6	2	2	2	2	3	0	1	2	3	1	0	0	0	0	0	0
A7	2	2	2	0	0	0	0	2	3	0	0	0	0	0	0	0
A8	0	0	0	1	2	2	2	0	2	0	0	0	0	0	0	0
A9	0	2	2	2	3	3	2	3	0	2	1	1	1	1	0	0
A10	0	3	2	2	1	1	2	3	2	0	0	0	0	2	0	0
A11	0	4	3	2	0	4	4	0	4	4	0	0	0	0	0	0
A12	1	2	3	2	1	2	1	1	3	1	3	0	1	1	2	2
A13	0	0	0	0	0	0	0	0	0	0	0	0	0	0	0	0
A14	0	2	1	2	2	1	1	3	3	2	0	0	0	0	1	0
A15	4	3	0	0	0	0	1	0	0	0	0	0	0	0	0	0
A16	1	3	3	3	3	3	3	3	3	2	1	2	2	2	0	0
A17	0	0	0	0	4	4	0	0	0	0	0	0	0	0	0	0
A18	0	4	3	2	3	2	2	3	1	1	0	0	0	0	0	0
A19	0	0	0	0	4	4	0	0	0	0	0	0	0	0	0	0
A20	1	3	2	4	3	4	3	3	3	1	2	2	2	2	2	0

(*continued*)

TABLE B.4 *(Continued)*

ID	A1	A2	A3	A4	A5	A6	A7	A8	A9	A10	A11	A12	A13	A14	A15	A16
A21	0	4	4	4	4	2	0	4	4	0	0	0	0	2	0	0
A22	0	4	3	4	4	4	3	4	4	4	0	0	0	3	0	0
A23	0	2	0	0	4	3	3	2	3	0	0	0	0	0	0	0
A24	0	3	0	2	3	0	0	0	4	0	0	0	0	0	0	0
A25	0	0	0	3	0	3	2	0	2	2	0	0	0	0	0	0
A26	0	2	2	2	2	1	2	2	2	2	0	0	0	0	0	0
A27	0	2	0	0	0	0	3	0	3	0	0	0	0	0	0	0
A28	0	1	0	3	0	3	0	0	3	0	0	0	0	0	0	0
A29	0	2	2	2	2	1	1	2	2	3	1	1	1	2	1	2
A30	1	3	3	4	3	3	3	3	3	3	2	2	2	2	2	3
A31	0	0	0	4	0	4	0	0	3	0	0	0	0	0	0	0
A32	0	1	1	2	2	1	2	0	2	1	2	0	0	2	0	0
A33	0	0	0	2	0	0	0	0	2	2	0	0	2	0	0	0
A34	0	0	0	0	0	0	0	0	0	0	0	0	0	0	0	0
A35	0	4	0	3	3	4	3	0	2	2	0	0	0	0	0	0
A36	0	4	0	0	4	4	4	0	4	0	0	0	0	0	0	0
A37	0	3	0	0	4	3	3	0	0	0	0	0	0	0	0	0
A38	0	4	0	4	4	3	0	3	3	3	0	0	0	3	3	0
A39	0	3	0	3	0	0	2	0	2	0	0	0	0	2	0	0
A40	0	0	0	2	4	4	2	2	4	1	0	0	0	0	0	0
A41	0	2	0	3	0	4	2	2	3	0	2	0	0	0	0	0
A42	3	0	0	0	0	0	0	0	0	0	0	0	0	0	0	0
A43	2	0	0	2	0	2	0	0	0	0	0	0	0	1	0	0

TABLE B.4   (*Continued*)

ID	A17	A18	A19	A20	A21	A22	A23	A24	A25	A26	A27	A28	A29	A30	A31
A1	0	0	0	0	0	0	0	0	0	0	0	0	0	0	0
A2	0	0	1	1	0	0	0	0	0	0	0	0	0	0	0
A3	0	0	0	0	0	0	0	0	0	0	0	0	0	0	0
A4	0	0	0	0	0	0	0	0	0	0	0	2	0	0	0
A5	0	0	0	0	0	0	0	0	0	0	0	0	0	0	0
A6	0	1	0	1	0	0	0	0	0	0	0	1	0	0	0
A7	0	0	0	0	0	0	0	0	0	0	0	0	0	0	0
A8	0	0	0	0	0	0	0	0	0	0	0	0	0	0	0
A9	1	1	1	3	1	0	0	0	0	1	0	2	1	0	0
A10	2	0	2	2	0	0	0	0	0	0	0	0	0	0	0
A11	0	0	0	0	0	0	0	0	0	0	0	0	0	0	0
A12	1	1	2	1	1	2	0	1	0	2	2	1	1	0	0
A13	0	0	0	0	0	0	0	0	0	0	0	0	0	0	0
A14	1	0	1	0	0	0	0	0	0	2	2	0	0	0	0
A15	0	0	0	0	0	0	0	0	0	0	0	0	0	0	0
A16	2	2	0	0	2	1	0	0	0	2	2	0	2	0	0
A17	0	0	0	0	0	0	0	0	0	0	0	0	0	0	0
A18	0	0	0	0	0	0	0	0	0	0	0	0	0	0	0
A19	0	0	0	0	0	0	0	0	0	0	0	0	0	0	0
A20	2	2	2	0	2	0	0	0	0	0	0	2	0	0	0

(*continued*)

TABLE B.4 (*Continued*)

ID	A17	A18	A19	A20	A21	A22	A23	A24	A25	A26	A27	A28	A29	A30	A31
A21	0	2	0	2	0	0	0	0	0	2	0	0	0	0	0
A22	3	0	3	0	0	0	0	0	0	3	3	0	0	0	0
A23	0	0	0	0	0	0	0	0	0	0	0	0	0	0	0
A24	0	0	0	0	0	0	0	0	0	0	0	0	0	0	0
A25	0	0	0	0	3	0	0	0	0	0	0	3	0	0	0
A26	2	0	0	0	0	0	0	0	0	0	0	0	0	0	0
A27	0	0	0	0	0	0	0	0	0	0	0	0	0	0	0
A28	0	2	2	0	0	0	0	0	0	0	0	0	0	0	0
A29	2	2	0	1	2	2	0	0	0	2	3	2	0	1	1
A30	2	3	2	2	3	3	2	2	2	3	3	3	3	0	2
A31	0	0	0	0	0	0	0	0	0	0	0	0	0	0	0
A32	1	2	2	1	0	0	0	0	0	1	2	0	0	0	0
A33	0	0	0	0	0	0	0	0	0	0	2	0	0	0	0
A34	0	0	0	0	0	0	0	0	0	0	0	0	0	0	0
A35	0	0	0	0	0	0	0	0	0	0	0	0	0	0	0
A36	3	0	0	0	0	0	0	0	0	0	0	0	0	0	0
A37	0	0	0	0	0	0	0	0	0	0	0	0	0	0	0
A38	0	2	2	3	3	0	0	0	0	3	0	3	0	0	0
A39	0	0	2	0	0	0	0	0	0	0	0	0	0	0	0
A40	2	2	0	0	0	0	0	0	0	0	0	0	0	0	0
A41	0	0	0	0	0	0	0	0	0	0	0	0	0	0	0
A42	0	0	0	0	0	0	0	0	0	0	0	0	0	0	0
A43	0	0	0	2	0	0	0	0	0	1	1	0	0	0	0

TABLE B.4   (*Continued*)

ID	A32	A33	A34	A35	A36	A37	A38	A39	A40	A41	A42	A43
A1	0	0	0	0	0	0	0	0	0	0	0	0
A2	0	0	0	0	0	0	0	0	1	1	0	0
A3	0	0	0	0	0	0	0	0	0	0	0	0
A4	0	0	0	0	0	0	0	0	0	0	0	0
A5	0	0	0	0	0	0	0	0	0	0	0	0
A6	0	0	0	0	1	1	0	0	2	1	0	0
A7	0	0	0	0	0	0	0	0	0	0	0	0
A8	0	0	0	0	0	0	0	0	0	0	0	0
A9	0	0	0	0	1	1	0	1	2	2	2	2
A10	0	0	0	0	0	0	0	0	0	0	0	0
A11	0	0	0	0	0	0	0	0	0	0	0	0
A12	0	0	0	0	0	1	0	1	1	1	1	1
A13	0	0	0	0	0	0	0	0	0	0	0	0
A14	0	0	0	0	0	0	0	0	1	1	0	0
A15	0	0	0	0	0	0	0	0	0	0	0	0
A16	0	0	0	0	0	0	0	0	0	0	0	0
A17	0	0	0	0	0	0	0	0	0	0	0	0
A18	0	0	0	0	0	0	0	0	0	0	0	0
A19	0	0	0	0	0	0	0	0	0	0	0	0
A20	0	0	0	0	2	0	0	2	3	2	2	2

(*continued*)

TABLE B.4 (*Continued*)

ID	A32	A33	A34	A35	A36	A37	A38	A39	A40	A41	A42	A43
A21	0	0	0	0	0	0	0	0	0	0	0	0
A22	0	0	0	0	0	0	0	0	0	0	0	0
A23	0	0	0	0	0	0	0	0	0	0	0	0
A24	0	0	0	0	0	0	0	0	0	0	0	0
A25	0	0	0	0	0	0	0	0	0	0	0	0
A26	0	0	0	0	0	0	0	0	0	0	0	0
A27	0	0	0	0	0	0	0	0	0	0	0	0
A28	1	0	0	0	0	0	0	0	0	0	0	0
A29	2	2	1	1	0	0	0	0	1	0	2	1
A30	2	2	1	1	2	2	2	2	3	2	2	2
A31	0	0	0	0	0	0	0	0	0	0	0	0
A32	0	0	0	0	0	0	0	0	0	0	0	0
A33	0	0	0	0	0	0	0	0	0	0	0	0
A34	0	0	0	0	0	0	0	0	0	0	0	0
A35	0	0	0	0	0	0	0	0	0	0	0	0
A36	0	0	0	0	0	0	0	0	3	0	0	0
A37	0	0	0	0	0	0	0	0	0	0	0	0
A38	0	0	0	0	0	0	0	0	2	3	3	2
A39	0	0	0	0	0	0	0	0	0	0	0	0
A40	0	0	0	0	0	0	0	0	0	0	0	0
A41	0	0	0	0	0	0	0	0	0	0	0	0
A42	0	0	0	0	0	0	0	0	0	0	0	0
A43	0	0	0	0	0	0	0	0	0	0	0	0

TABLE B.5   Health Organization — Agent × Agent Clinical Advice

ID	A1	A2	A3	A4	A5	A6	A7	A8	A9	A10	A11	A12	A13	A14	A15	A16
A1	0	1	2	0	3	0	0	0	0	0	0	0	0	0	0	0
A2	3	0	2	0	2	0	0	1	2	0	0	0	0	0	0	0
A3	4	0	0	2	0	2	0	0	0	0	0	0	0	0	0	0
A4	4	0	0	0	0	0	0	0	0	0	0	0	0	0	0	0
A5	4	0	2	0	0	2	0	0	0	0	0	0	0	0	0	0
A6	4	0	2	0	0	0	1	0	0	0	0	0	0	0	0	0
A7	4	0	3	0	0	0	0	0	0	0	0	0	0	0	0	0
A8	2	0	0	0	0	0	0	0	0	0	0	0	0	0	0	0
A9	3	2	3	2	1	3	2	3	0	1	0	0	0	1	0	0
A10	3	2	3	0	0	0	1	0	0	0	0	0	0	0	0	0
A11	4	0	4	0	0	0	0	0	0	0	0	0	0	0	0	0
A12	1	3	2	2	3	2	1	2	3	2	3	0	0	2	1	0
A13	3	3	3	2	3	2	2	2	3	2	1	1	0	2	0	1
A14	4	0	3	0	0	0	0	0	0	0	0	0	0	0	0	0
A15	4	3	1	0	0	0	0	0	0	0	0	0	0	0	0	0
A16	4	1	3	1	1	1	1	1	1	1	0	0	0	0	0	0
A17	4	0	3	0	0	0	0	0	0	0	0	0	0	0	0	0
A18	4	0	2	0	0	0	0	0	0	0	0	0	0	0	0	0
A19	4	0	0	0	0	0	0	0	0	0	0	0	0	0	0	0
A20	3	3	3	2	2	3	2	3	3	1	0	0	0	0	0	0

(*continued*)

TABLE B.5 (*Continued*)

ID	A1	A2	A3	A4	A5	A6	A7	A8	A9	A10	A11	A12	A13	A14	A15	A16
A21	2	4	4	0	4	0	4	4	0	0	0	0	0	0	0	0
A22	4	0	3	0	0	0	0	0	0	0	0	0	0	0	0	0
A23	4	0	3	0	0	0	0	0	0	0	0	0	0	0	0	0
A24	4	0	1	0	0	0	0	0	0	0	0	0	0	0	0	0
A25	2	0	0	2	0	2	2	0	0	0	0	0	0	0	0	0
A26	4	0	3	0	0	0	2	0	0	0	0	0	0	0	0	0
A27	2	0	2	0	0	0	0	0	0	0	0	0	0	0	0	0
A28	1	0	2	0	0	0	0	0	0	0	0	0	0	0	0	0
A29	1	0	2	0	0	0	0	0	0	0	0	0	0	0	0	0
A30	2	1	3	2	1	2	1	1	1	1	1	1	1	1	1	1
A31	4	0	0	0	0	0	0	0	0	0	0	0	0	0	0	0
A32	2	1	0	2	2	1	2	0	1	0	0	0	0	0	0	0
A33	4	0	0	0	0	3	0	0	0	0	0	0	0	0	0	0
A34	0	0	0	0	0	0	0	0	0	0	0	0	0	0	0	0
A35	4	0	2	0	0	0	0	0	0	0	0	0	0	0	0	0
A36	4	0	2	0	0	0	0	0	0	0	0	0	0	0	0	0
A37	4	0	0	0	0	0	0	0	0	0	0	0	0	0	0	0
A38	4	0	2	0	0	0	0	0	0	0	0	0	0	0	0	0
A39	0	2	0	0	0	0	2	0	2	0	0	0	0	0	0	0
A40	4	0	0	0	0	0	0	0	0	0	0	0	0	0	0	0
A41	4	0	2	0	0	2	0	0	0	0	0	0	0	0	0	0
A42	3	0	0	0	0	0	0	0	0	0	0	0	0	0	0	0
A43	3	0	0	0	0	0	0	0	0	0	0	0	0	0	0	0

TABLE B.5   *(Continued)*

ID	A17	A18	A19	A20	A21	A22	A23	A24	A25	A26	A27	A28	A29	A30	A31
A1	0	0	0	0	0	0	0	0	0	0	0	0	0	0	0
A2	0	0	0	0	0	0	0	0	0	0	0	0	0	0	0
A3	0	0	0	0	0	0	0	0	0	0	0	0	0	0	0
A4	0	0	0	0	0	0	0	0	0	0	0	0	0	0	0
A5	0	0	0	0	0	0	0	0	0	0	0	0	0	0	0
A6	0	0	0	0	0	0	0	0	0	0	0	0	0	0	0
A7	0	0	0	0	0	0	0	0	0	0	0	0	0	0	0
A8	0	0	0	0	0	0	0	0	0	0	0	0	0	0	0
A9	0	0	0	0	0	0	0	0	0	0	0	0	0	0	0
A10	0	0	0	0	0	0	0	0	0	0	0	0	0	0	0
A11	0	0	0	0	0	0	0	0	0	0	0	0	0	0	0
A12	2	1	3	1	0	1	0	3	0	1	1	2	1	0	0
A13	2	2	1	1	2	2	1	2	0	2	2	1	1	0	0
A14	0	0	0	0	0	0	0	0	0	0	0	0	0	0	0
A15	0	0	0	0	0	0	0	0	0	0	0	0	0	0	0
A16	1	0	0	0	0	0	0	0	0	0	1	0	0	0	0
A17	0	0	0	0	0	0	0	0	0	0	0	0	0	0	0
A18	0	0	0	0	0	0	0	0	0	0	0	0	0	0	0
A19	0	0	0	0	0	0	0	0	0	0	0	0	0	0	0
A20	0	0	2	0	0	0	0	0	0	0	0	0	0	0	0

(continued)

TABLE B.5 (*Continued*)

ID	A17	A18	A19	A20	A21	A22	A23	A24	A25	A26	A27	A28	A29	A30	A31
A21	0	0	4	0	0	0	0	0	0	0	0	0	0	0	0
A22	0	0	0	0	0	0	0	0	0	0	0	0	0	0	0
A23	0	0	0	0	0	0	0	0	0	0	0	0	0	0	0
A24	0	0	0	0	0	0	0	0	0	0	0	0	0	0	0
A25	0	0	0	0	0	0	0	0	0	0	0	0	0	0	0
A26	0	0	0	0	0	0	0	0	0	0	0	0	0	0	0
A27	0	0	0	0	0	0	0	0	0	0	0	0	0	0	0
A28	0	0	0	0	0	0	0	0	0	0	0	0	0	0	0
A29	0	0	0	0	0	0	0	0	0	0	0	0	0	0	0
A30	1	1	1	1	1	1	1	1	1	1	1	1	1	0	1
A31	0	0	0	0	0	0	0	0	0	0	0	0	0	0	0
A32	0	0	0	0	0	0	0	0	0	0	0	0	0	0	0
A33	0	0	0	0	0	0	0	0	0	0	0	0	0	0	0
A34	0	0	0	0	0	0	0	0	0	0	0	0	0	0	0
A35	0	0	0	0	0	0	0	0	0	0	0	0	0	0	0
A36	0	0	0	0	0	0	0	0	0	0	0	0	0	0	0
A37	0	0	0	0	0	0	0	0	0	0	0	0	0	0	0
A38	0	0	0	0	0	0	0	0	0	0	0	0	0	0	0
A39	0	0	2	0	0	0	0	0	0	0	0	0	0	0	0
A40	0	0	0	0	0	0	0	0	0	0	0	0	0	0	0
A41	0	0	0	0	0	0	0	0	0	0	0	0	0	0	0
A42	0	0	0	0	0	0	0	0	0	0	0	0	0	0	0
A43	0	0	0	0	0	0	0	0	0	0	0	0	0	0	0

TABLE B.5   (*Continued*)

ID	A32	A33	A34	A35	A36	A37	A38	A39	A40	A41	A42	A43
A1	0	0	0	0	0	0	0	0	0	0	0	0
A2	0	0	0	0	0	0	0	0	0	0	0	0
A3	0	0	0	0	0	0	0	0	0	0	0	0
A4	0	0	0	0	0	0	0	0	0	0	0	0
A5	0	0	0	0	0	0	0	0	0	0	0	0
A6	0	0	0	0	0	0	0	0	0	0	0	0
A7	0	0	0	0	0	0	0	0	0	0	0	0
A8	0	0	0	0	0	0	0	0	0	0	0	0
A9	0	0	0	0	2	0	0	1	2	2	2	2
A10	0	0	0	0	0	0	0	0	0	0	0	0
A11	0	0	0	0	0	0	0	0	0	0	0	0
A12	0	0	0	0	1	0	0	1	1	1	0	0
A13	0	0	0	0	0	0	0	0	2	1	1	0
A14	0	0	0	0	0	0	0	0	0	0	0	0
A15	0	0	0	0	0	0	0	0	0	0	0	0
A16	0	0	0	0	0	0	0	0	0	0	0	0
A17	0	0	0	0	0	0	0	0	0	0	0	0
A18	0	0	0	0	0	0	0	0	0	0	0	0
A19	0	0	0	0	0	0	0	0	0	0	0	0
A20	0	0	0	0	0	0	0	0	3	2	0	0

(*continued*)

TABLE B.5 (*Continued*)

ID	A32	A33	A34	A35	A36	A37	A38	A39	A40	A41	A42	A43
A21	0	0	0	0	2	0	0	0	2	0	0	0
A22	0	0	0	0	0	0	0	0	0	0	0	0
A23	0	0	0	0	0	0	0	0	0	0	0	0
A24	0	0	0	0	0	0	0	0	0	0	0	0
A25	0	0	0	0	0	0	0	0	0	0	0	0
A26	0	0	0	0	0	0	0	0	0	0	0	0
A27	0	0	0	0	0	0	0	0	0	0	0	0
A28	0	0	0	0	0	0	0	0	0	0	0	0
A29	0	0	0	0	0	0	0	0	0	0	0	0
A30	1	1	1	1	1	1	1	1	1	1	1	1
A31	0	0	0	0	0	0	0	0	0	0	0	0
A32	0	0	0	0	0	0	0	0	0	0	0	0
A33	0	0	0	0	0	0	0	0	0	0	0	0
A34	0	0	0	0	0	0	0	0	0	0	0	0
A35	0	0	0	0	0	0	0	0	0	0	0	0
A36	0	0	0	0	0	0	0	0	0	0	0	0
A37	0	0	0	0	0	0	0	0	0	0	0	0
A38	0	0	0	0	0	0	0	0	0	0	0	0
A39	0	0	0	0	0	0	0	0	0	0	0	0
A40	0	0	0	0	0	0	0	0	0	0	0	0
A41	0	0	0	0	0	0	0	0	0	0	0	0
A42	0	0	0	0	0	0	0	0	0	0	0	0
A43	0	0	0	0	0	0	0	0	0	0	0	0

TABLE B.6   Dolphin Agent Names

Agent ID No.	Title
0	Beak
1	Beescratch
2	Bumper
3	CCL
4	Cross
5	DN16
6	DN21
7	DN63
8	Double
9	Feather
10	Fish
11	Five
12	Fork
13	Gallatin
14	Grin
15	Haecksel
16	Hook
17	Jet
18	Jonah
19	Knit
20	Kringel
21	MN105
22	MN23
23	MN60
24	MN83
25	Mus
26	Notch
27	Number1
28	Oscar
29	Patchback
30	PL

(*continued*)

TABLE B.6 (*Continued*)

Agent ID No.	Title
31	Quasi
32	Ripplefluke
33	Scabs
34	Shmuddel
35	SMN5
36	SN100
37	SN4
38	SN63
39	SN89
40	SN9
41	SN90
42	SN96
43	Stripes
44	Thumper
45	Topless
46	TR120
47	TR77
48	TR82
49	TR88
50	TR99
51	Trigger
52	TSN103
53	TSN83
54	Upbang
55	Vau
56	Wave
57	Web
58	Whitetip
59	Zap
60	Zig
61	Zipfel

TABLE B.7   Dolphins Agent × Agent Network

Dolphin	0	1	2	3	4	5	6	7	8	9	10	11	12	13	14	15	16	17	18	19	20
0	0	0	0	0	0	0	0	0	0	0	0	1	0	0	0	1	1	0	0	0	0
1	0	0	0	0	0	0	0	0	0	0	0	0	0	0	0	0	0	1	0	1	0
2	0	0	0	0	0	0	0	0	0	0	1	0	0	0	0	0	0	0	0	0	0
3	0	0	0	0	0	0	0	0	1	0	0	0	0	0	1	0	0	0	0	0	0
4	0	0	0	0	0	0	0	0	0	0	0	0	0	0	0	0	0	0	0	0	0
5	0	0	0	0	0	0	0	0	0	1	0	0	0	1	0	0	0	0	0	0	0
6	0	0	0	0	0	0	0	0	0	1	0	0	0	1	0	0	0	1	0	0	0
7	0	0	0	0	0	0	0	0	0	0	0	0	0	0	0	0	0	0	0	1	0
8	0	0	0	1	0	0	0	0	0	0	0	0	0	0	0	0	0	0	0	0	1
9	0	0	0	0	0	1	1	0	0	0	0	0	0	1	0	0	0	1	0	0	0
10	1	0	1	0	0	0	0	0	0	0	0	0	0	0	0	0	0	0	0	0	0
11	0	0	0	0	0	0	0	0	0	0	0	0	0	0	0	0	0	0	0	0	0
12	0	0	0	0	0	0	0	0	0	0	0	0	0	0	0	0	0	0	0	0	0
13	0	0	0	0	0	1	1	0	0	1	0	0	0	0	0	0	0	1	0	0	0
14	1	0	0	1	0	0	0	0	0	0	0	0	0	0	0	0	1	0	0	0	0
15	1	0	0	0	0	0	0	0	0	0	0	0	0	0	0	0	0	0	1	0	0
16	0	0	0	0	0	0	0	0	0	0	0	0	0	0	1	0	0	0	0	0	1
17	0	1	0	0	0	0	1	0	0	1	0	0	0	1	0	0	0	0	0	0	0
18	0	0	0	0	0	0	0	0	0	0	0	0	0	0	0	1	0	0	0	0	1
19	0	1	0	0	0	0	0	1	0	0	0	0	0	0	0	0	0	0	0	0	0
20	0	0	0	0	0	0	0	0	1	0	0	0	0	0	0	0	1	0	1	0	0
21	0	0	0	0	0	0	0	0	0	0	0	0	0	0	0	0	0	0	1	0	0
22	0	0	0	0	0	0	0	0	0	0	0	0	0	0	0	0	0	1	0	0	0
23	0	0	0	0	0	0	0	0	0	0	0	0	0	0	0	0	0	0	0	0	0
24	0	0	0	0	0	0	0	0	0	0	0	0	0	0	1	1	0	0	1	0	0
25	0	0	0	0	0	0	0	0	0	0	0	0	0	0	0	0	0	0	1	0	0
26	0	1	0	0	0	0	0	0	0	0	0	0	0	0	0	0	0	0	0	0	0
27	0	1	0	0	0	0	0	1	0	0	0	0	0	0	0	0	0	1	0	0	0
28	0	1	0	0	0	0	0	0	1	0	0	0	0	0	0	0	0	0	0	0	1
29	0	0	0	0	0	0	0	0	0	0	1	0	0	0	0	0	0	0	1	0	0
30	0	0	0	0	0	0	0	1	0	0	0	0	0	0	0	0	0	0	0	1	0

(continued)

TABLE B.7   (*Continued*)

Dolphin	0	1	2	3	4	5	6	7	8	9	10	11	12	13	14	15	16	17	18	19	20
31	0	0	0	0	0	0	0	0	0	0	0	0	0	0	0	0	0	1	0	0	0
32	0	0	0	0	0	0	0	0	0	1	0	0	0	1	0	0	0	0	0	0	0
33	0	0	0	0	0	0	0	0	0	0	0	0	1	0	1	0	1	0	0	0	0
34	0	0	0	0	0	0	0	0	0	0	0	0	0	0	1	0	0	0	0	0	0
35	0	0	0	0	0	0	0	0	0	0	0	0	0	0	0	0	0	0	0	0	0
36	0	1	0	0	0	0	0	0	0	0	0	0	0	0	0	0	0	0	0	0	1
37	0	0	0	0	0	0	0	0	1	0	0	0	0	0	1	0	1	0	0	0	0
38	0	0	0	0	0	0	0	0	0	0	0	0	0	0	1	0	1	0	0	0	1
39	0	0	0	0	0	0	0	0	0	0	0	0	0	0	0	0	0	0	0	0	0
40	1	0	0	0	0	0	0	1	0	0	0	0	0	0	1	1	0	0	0	0	0
41	0	1	0	0	0	0	0	0	0	0	1	0	0	0	1	0	0	0	0	0	0
42	1	0	1	0	0	0	0	0	0	0	1	0	0	0	0	0	0	0	0	0	0
43	0	0	0	0	0	0	0	0	0	0	0	0	0	0	1	0	0	0	0	0	0
44	0	0	1	0	0	0	0	0	0	0	0	0	0	0	0	0	0	0	0	0	1
45	0	0	0	0	0	0	0	0	1	0	0	0	0	0	0	1	0	0	1	0	0
46	0	0	0	0	0	0	0	0	0	0	0	0	0	0	0	0	0	0	0	0	0
47	1	0	0	0	0	0	0	0	0	0	0	1	0	0	0	0	0	0	0	0	1
48	0	0	0	0	0	0	0	0	0	0	0	0	0	0	0	0	0	0	0	0	0
49	0	0	0	0	0	0	0	0	0	0	0	0	0	0	0	0	0	0	0	0	0
50	0	0	0	0	0	0	0	0	0	0	0	0	0	0	1	0	1	0	0	0	1
51	0	0	0	0	1	0	0	0	0	0	0	1	0	0	0	0	0	0	1	0	0
52	0	0	0	0	0	0	0	0	0	0	0	0	0	0	1	0	0	0	0	0	0
53	0	0	0	0	0	0	0	0	0	0	0	0	0	0	0	0	0	0	0	0	0
54	0	1	0	0	0	0	1	1	0	0	0	0	0	1	0	0	0	0	0	1	0
55	0	0	0	0	0	0	0	0	0	0	0	0	0	0	0	1	0	0	0	0	0
56	0	0	0	0	0	1	1	0	0	0	0	0	0	0	0	0	0	0	0	0	0
57	0	0	0	0	0	1	1	0	0	1	0	0	0	1	0	0	0	1	0	0	0
58	0	0	0	0	0	0	0	0	0	0	0	0	0	0	0	0	0	0	0	0	0
59	0	0	0	1	0	0	0	0	1	0	0	0	0	0	0	1	0	0	0	0	0
60	0	0	0	0	0	0	0	0	0	0	0	0	0	0	0	0	0	0	0	0	0
61	0	0	1	0	0	0	0	0	0	0	0	0	0	0	0	0	0	0	0	0	0

TABLE B.7   (*Continued*)

Dolphin	21	22	23	24	25	26	27	28	29	30	31	32	33	34	35	36	37	38	39	40
0	0	0	0	0	0	0	0	0	0	0	0	0	0	0	0	0	0	0	0	1
1	0	0	0	0	0	1	1	1	0	0	0	0	0	0	0	0	1	0	0	0
2	0	0	0	0	0	0	0	0	0	0	0	0	0	0	0	0	0	0	0	0
3	0	0	0	0	0	0	0	0	0	0	0	0	0	0	0	0	0	0	0	0
4	0	0	0	0	0	0	0	0	0	0	0	0	0	0	0	0	0	0	0	0
5	0	0	0	0	0	0	0	0	0	0	0	0	0	0	0	0	0	0	0	0
6	0	0	0	0	0	0	0	0	0	0	0	0	0	0	0	0	0	0	0	0
7	0	0	0	0	0	0	1	0	0	1	0	0	0	0	0	0	0	0	0	1
8	0	0	0	0	0	0	0	1	0	0	0	0	0	0	0	0	1	0	0	0
9	0	0	0	0	0	0	0	0	0	0	1	0	0	0	0	0	0	0	0	0
10	0	0	0	0	0	0	0	0	1	0	0	0	0	0	0	0	0	0	0	0
11	0	0	0	0	0	0	0	0	0	0	0	0	0	0	0	0	0	0	0	0
12	0	0	0	0	0	0	0	0	0	0	0	0	1	0	0	0	0	0	0	0
13	0	0	0	0	0	0	0	0	0	0	0	0	1	0	0	0	0	0	0	0
14	0	0	0	1	0	0	0	0	0	0	0	0	0	1	1	0	0	1	1	1
15	0	0	0	1	0	0	0	0	0	0	0	0	0	0	0	0	0	0	0	1
16	0	0	0	0	0	0	0	0	0	0	0	0	0	1	0	0	0	1	1	0
17	0	1	0	0	1	0	1	0	0	0	1	0	0	0	0	0	0	0	0	0
18	1	0	0	1	0	0	0	0	1	0	0	0	0	0	0	0	0	0	0	0
19	0	0	0	0	0	0	0	0	0	1	0	0	0	0	0	0	0	0	0	0
20	0	0	0	0	0	0	0	1	0	0	0	0	0	0	0	1	0	1	0	0
21	0	0	0	0	0	0	0	0	1	0	0	0	1	0	0	0	1	0	0	0
22	0	0	0	0	0	0	0	0	0	0	0	0	0	0	0	0	0	0	0	0
23	0	0	0	0	0	0	0	0	0	0	0	0	0	0	0	1	0	0	0	0
24	0	0	0	0	0	0	0	0	1	0	0	0	0	0	0	0	0	0	0	0
25	0	0	0	0	0	1	1	0	0	0	0	0	0	0	0	0	0	0	0	0
26	0	0	0	0	1	0	1	0	0	0	0	0	0	0	0	0	0	0	0	0
27	0	0	0	0	1	1	0	0	0	0	0	0	0	0	0	0	0	0	0	0
28	0	0	0	0	0	0	0	0	0	1	0	0	0	0	0	0	0	0	0	0
29	1	0	0	1	0	0	0	0	0	0	0	0	0	0	1	0	0	0	0	0
30	0	0	0	0	0	0	0	1	0	0	0	0	0	0	0	0	0	0	0	0

(*continued*)

TABLE B.7 (*Continued*)

Dolphin	21	22	23	24	25	26	27	28	29	30	31	32	33	34	35	36	37	38	39	40
31	0	0	0	0	0	0	0	0	0	0	0	0	0	0	0	0	0	0	0	0
32	0	0	0	0	0	0	0	0	0	0	0	0	0	0	0	0	0	0	0	0
33	1	0	0	0	0	0	0	0	0	0	0	0	0	1	0	0	1	1	0	1
34	0	0	0	0	0	0	0	0	0	0	0	0	1	0	0	0	1	1	0	0
35	0	0	0	0	0	0	0	0	0	1	0	0	0	0	0	0	0	0	0	0
36	0	0	1	0	0	0	0	0	0	0	0	0	0	0	0	0	1	0	1	1
37	1	0	0	0	0	0	0	0	0	0	0	0	1	1	0	1	0	0	0	1
38	0	0	0	0	0	0	0	0	0	0	0	0	0	1	0	0	0	0	0	0
39	0	0	0	0	0	0	0	0	0	0	0	0	0	0	0	1	0	0	0	0
40	0	0	0	0	0	0	0	0	0	0	0	0	1	0	0	1	1	0	0	0
41	0	0	0	0	0	0	0	0	0	0	0	0	0	0	0	0	0	0	0	0
42	0	0	0	0	0	0	0	0	0	1	0	0	0	0	0	0	0	0	0	0
43	0	0	0	0	0	0	0	0	0	1	0	0	0	1	0	0	0	1	1	0
44	0	0	0	0	0	0	0	0	0	0	0	0	0	1	0	0	0	1	0	0
45	1	0	1	1	0	0	0	0	0	1	0	0	0	0	0	0	0	1	0	0
46	0	0	0	0	0	0	0	0	0	0	0	0	0	0	0	0	0	0	0	0
47	0	0	0	0	0	0	0	1	0	1	0	0	0	0	0	0	0	0	0	0
48	0	0	0	0	0	0	0	0	0	0	0	0	0	0	0	0	0	0	0	0
49	0	0	0	0	0	0	0	0	0	0	0	0	0	1	0	0	0	0	0	0
50	0	0	0	0	0	0	0	0	0	0	0	0	1	0	0	0	0	0	0	0
51	1	0	1	1	0	0	0	0	1	0	0	0	0	0	0	0	0	0	0	0
52	0	0	0	0	0	0	0	0	1	0	0	0	0	0	0	0	0	1	0	1
53	0	0	0	0	0	0	0	0	0	0	0	0	0	0	0	0	0	0	0	0
54	0	0	0	0	0	0	0	0	0	0	0	0	0	0	0	0	0	0	0	0
55	0	0	0	0	0	0	0	0	0	0	0	0	0	0	0	0	0	0	0	0
56	0	0	0	0	0	0	0	0	0	0	0	0	0	0	0	0	0	0	0	0
57	0	0	0	0	0	0	0	0	0	0	0	0	0	0	0	0	0	0	1	0
58	0	0	0	0	0	0	0	0	0	0	0	0	0	0	0	0	0	1	0	0
59	0	0	0	0	0	0	0	0	0	0	0	0	0	0	0	1	0	0	0	0
60	0	0	0	0	0	0	0	0	0	0	0	1	0	0	0	0	0	0	0	0
61	0	0	0	0	0	0	0	0	0	0	0	0	0	0	0	0	1	0	0	0

TABLE B.7   (*Continued*)

Dolphin	41	42	43	44	45	46	47	48	49	50	51	52	53	54	55	56	57	58	59	60	61
0	0	1	0	0	0	0	1	0	0	0	0	0	0	0	0	0	0	0	0	0	0
1	1	0	0	0	0	0	0	0	0	0	0	0	0	0	1	0	0	0	0	0	0
2	0	1	0	1	0	0	0	0	0	0	0	0	0	0	0	0	0	0	0	0	1
3	0	0	0	0	0	0	0	0	0	0	0	0	0	0	0	0	0	0	1	0	0
4	0	0	0	0	0	0	0	0	0	0	0	1	0	0	0	0	0	0	0	0	0
5	0	0	0	0	0	0	0	0	0	0	0	0	0	0	0	0	1	1	0	0	0
6	0	0	0	0	0	0	0	0	0	0	0	0	0	1	0	1	1	0	0	0	0
7	0	0	0	0	0	0	0	0	0	0	0	0	0	0	1	0	0	0	0	0	0
8	0	0	0	0	1	0	0	0	0	0	0	0	0	0	0	0	0	0	1	0	0
9	1	0	0	0	0	0	0	0	0	0	0	0	0	0	0	0	1	0	0	0	0
10	0	1	0	0	0	0	1	0	0	0	0	0	0	0	0	0	0	0	0	0	0
11	0	0	0	0	0	0	0	0	0	0	1	0	0	0	0	0	0	0	0	0	0
12	0	0	0	0	0	0	0	0	0	0	0	0	0	0	0	0	0	0	0	0	0
13	1	0	0	0	0	0	0	0	0	0	0	0	0	1	0	0	1	0	0	0	0
14	0	0	1	0	0	0	0	0	0	1	0	1	0	0	0	0	0	0	0	0	0
15	0	0	0	0	1	0	0	0	0	0	0	0	0	0	0	1	0	0	0	1	0
16	0	0	0	0	0	0	0	0	0	1	0	0	0	0	0	0	0	0	0	0	0
17	0	0	0	0	0	0	0	0	0	0	0	0	0	0	0	0	0	1	0	0	0
18	0	0	0	0	1	0	0	0	0	0	1	0	0	0	0	0	0	0	0	0	0
19	0	0	0	0	0	0	0	0	0	0	0	0	0	1	0	0	0	0	0	0	0
20	0	0	0	1	0	0	1	0	0	1	0	0	0	0	0	0	0	0	0	0	0
21	0	0	0	0	1	0	0	0	0	0	1	0	0	0	0	0	0	0	0	0	0
22	0	0	0	0	0	0	0	0	0	0	0	0	0	0	0	0	0	0	0	0	0
23	0	0	0	0	1	0	0	0	0	0	1	0	0	0	0	0	0	0	0	0	0
24	0	0	0	0	1	0	0	0	0	0	1	0	0	0	0	0	0	0	0	0	0
25	0	0	0	0	0	0	0	0	0	0	0	0	0	0	0	0	0	0	0	0	0
26	0	0	0	0	0	0	0	0	0	0	0	0	0	0	0	0	0	0	0	0	0
27	0	0	0	0	0	0	0	0	0	0	0	0	0	0	0	0	0	0	0	0	0
28	0	0	0	0	0	0	1	0	0	0	0	0	0	0	0	0	0	0	0	0	0
29	0	0	1	0	1	0	0	0	0	0	1	1	0	0	0	0	0	0	0	0	0
30	0	1	0	0	0	0	1	0	0	0	0	0	0	0	0	0	0	0	0	0	0

(*continued*)

TABLE B.7 (*Continued*)

Dolphin	41	42	43	44	45	46	47	48	49	50	51	52	53	54	55	56	57	58	59	60	61
31	0	0	0	0	0	0	0	0	0	0	0	0	0	0	0	0	0	0	0	0	0
32	0	0	0	0	0	0	0	0	0	0	0	0	0	0	0	0	0	0	0	1	0
33	0	0	1	0	0	0	0	0	0	0	1	0	0	0	0	0	0	0	0	0	0
34	0	0	0	1	0	0	0	0	1	0	0	0	0	0	0	0	0	0	0	0	0
35	0	0	0	0	0	0	0	0	0	0	0	0	0	0	0	0	0	0	0	0	0
36	0	0	0	0	0	0	0	0	0	0	0	0	0	0	0	0	0	0	1	0	0
37	0	0	1	0	1	0	0	0	0	0	0	0	0	0	0	0	0	0	0	0	1
38	0	0	1	1	0	0	0	0	0	0	0	0	1	0	0	0	0	0	1	0	0
39	0	0	0	0	0	0	0	0	0	0	0	0	0	0	0	0	1	0	0	0	0
40	0	0	0	0	0	0	0	0	0	0	0	1	0	0	0	0	0	0	0	0	0
41	0	0	0	0	0	0	0	0	0	0	0	0	0	1	0	0	1	0	0	0	0
42	0	0	0	0	0	0	1	0	0	1	0	0	0	0	0	0	0	0	0	0	0
43	0	0	0	0	0	1	0	0	0	0	0	0	0	1	0	0	0	0	0	0	0
44	0	0	0	0	0	0	0	0	0	0	0	0	0	0	0	0	0	0	0	0	0
45	0	0	0	0	0	0	0	0	0	1	1	0	0	0	0	0	0	0	1	0	0
46	0	0	1	0	0	0	0	0	1	0	0	0	0	0	0	0	0	0	0	0	0
47	0	1	0	0	0	0	0	0	0	0	0	0	0	0	0	0	0	0	0	0	0
48	0	0	0	0	0	0	0	0	0	0	0	0	0	0	0	0	1	0	0	0	0
49	0	0	0	0	0	1	0	0	0	0	0	0	0	0	0	0	0	0	0	0	0
50	0	1	0	0	1	0	0	0	0	0	1	0	0	0	0	0	0	0	0	0	0
51	0	0	0	0	1	0	0	0	0	1	0	0	0	0	1	0	0	0	0	0	0
52	0	0	0	0	0	0	0	0	0	0	0	0	0	0	0	0	0	0	0	0	0
53	0	0	1	0	0	0	0	0	0	0	0	0	0	0	0	0	0	0	0	0	1
54	1	0	0	0	0	0	0	0	0	0	0	0	0	0	0	0	1	0	0	0	0
55	0	0	0	0	0	0	0	0	0	0	1	0	0	0	0	0	0	0	0	0	0
56	0	0	0	0	0	0	0	0	0	0	0	0	0	0	0	0	0	0	0	0	0
57	1	0	0	0	0	0	0	1	0	0	0	0	0	1	0	0	0	0	0	0	0
58	0	0	0	0	0	0	0	0	0	0	0	0	0	0	0	0	0	0	0	0	0
59	0	0	0	0	1	0	0	0	0	0	0	0	0	0	0	0	0	0	0	0	0
60	0	0	0	0	0	0	0	0	0	0	0	0	0	0	0	0	0	0	0	0	0
61	0	0	0	0	0	0	0	0	0	0	0	0	1	0	0	0	0	0	0	0	0

TABLE B.8   Lab Exercise 5–Subgroup1 Agent × Agent Network

ID	A1	A2	A3	A4	A5	A6	A7	A8	A9	A10	A11	A12	A13	A14	A15	A16	A17	A18	A19	A20
A1	0	0	0	0	0	0	0	0	0	0	0	0	0	0	0	0	0	0	0	0
A2	0	0	0	0	0	0	0	1	0	0	0	0	0	0	0	0	0	0	0	0
A3	1	0	0	0	0	0	0	0	0	0	0	0	0	0	1	0	1	0	0	0
A4	1	0	0	0	0	0	0	0	0	0	0	0	0	1	0	0	0	0	0	0
A5	0	0	0	0	0	1	0	0	0	0	0	0	0	0	0	0	0	0	0	0
A6	0	0	0	0	1	0	0	0	0	0	0	0	0	0	0	0	0	0	0	0
A7	0	0	0	0	0	0	0	0	0	0	0	0	0	0	0	0	0	0	0	0
A8	0	1	0	0	0	0	0	0	0	0	0	0	0	0	0	0	0	0	0	0
A9	0	0	0	0	0	0	0	0	0	0	0	0	0	0	0	0	0	0	0	0
A10	0	0	0	0	0	0	0	0	0	0	0	0	0	1	0	0	0	0	1	0
A11	0	0	0	0	0	0	0	0	0	0	0	0	0	0	0	0	0	0	0	0
A12	0	0	0	0	1	0	0	0	0	0	0	0	0	0	0	0	0	0	0	0
A13	0	0	0	0	0	0	0	0	0	0	0	0	0	0	0	0	0	0	0	0
A14	0	0	1	0	0	0	0	0	0	0	0	0	0	0	0	0	0	0	0	0
A15	0	0	0	0	0	0	0	0	1	0	0	0	0	0	0	0	1	0	0	0
A16	1	0	1	0	0	0	0	0	0	0	0	0	0	1	0	0	0	0	0	0
A17	0	0	0	0	0	0	0	0	0	0	0	0	0	0	1	0	0	0	0	0
A18	0	0	0	0	0	0	0	0	0	1	0	0	1	0	0	0	0	0	0	0
A19	0	0	0	0	0	0	0	1	0	0	0	0	0	0	0	0	0	0	0	0
A20	0	0	0	0	0	0	0	0	0	0	0	0	0	0	0	0	0	1	0	0
A21	0	1	0	0	0	0	0	0	0	0	0	0	0	0	0	0	0	0	0	0
A22	0	0	0	0	0	0	0	0	0	0	0	0	0	0	0	0	0	0	0	0
A23	0	0	0	0	0	0	0	0	0	0	0	0	0	0	0	0	0	0	0	0
A24	1	0	1	1	0	0	0	0	0	0	0	0	0	0	0	1	0	0	0	0
A25	0	0	0	0	0	0	0	0	0	0	0	0	0	0	0	0	0	0	0	0
A26	0	0	0	0	0	0	0	0	0	0	0	0	0	0	0	0	0	0	0	0
A27	0	0	0	0	0	0	0	0	0	0	0	0	0	0	1	0	1	0	0	0
A28	0	0	0	0	1	0	0	0	0	0	0	0	0	0	0	0	0	0	0	0
A29	0	0	0	0	0	0	0	0	0	0	0	0	0	0	0	0	0	0	0	0
A30	0	0	1	1	0	0	0	0	0	0	0	0	0	0	0	0	0	0	0	0
A31	0	0	0	0	0	0	1	0	0	0	0	0	1	1	0	0	1	0	0	0
A32	0	0	0	0	0	0	0	1	0	0	0	0	0	0	0	0	0	0	0	0
A33	0	0	0	0	1	0	0	0	0	0	0	0	0	0	0	0	0	0	0	0
A34	0	0	0	0	0	0	0	1	0	0	0	0	0	0	0	0	0	0	1	0
A35	0	0	0	0	0	0	0	0	0	0	0	0	1	0	0	0	0	0	0	0
A36	0	0	0	0	0	1	0	0	0	0	0	0	0	0	0	0	0	0	0	0
A37	0	0	0	0	0	0	0	0	0	0	0	0	0	0	0	0	0	0	0	0
A38	0	0	0	0	0	0	0	0	0	0	0	0	0	0	0	0	0	0	0	0
A39	0	0	0	0	0	0	0	0	0	0	1	0	0	0	0	0	0	0	0	0

(*continued*)

TABLE B.8 (*Continued*)

ID	A1	A2	A3	A4	A5	A6	A7	A8	A9	A10	A11	A12	A13	A14	A15	A16	A17	A18	A19	A20
A40	0	0	0	1	0	0	0	0	0	0	0	0	0	0	0	0	1	0	0	0
A41	0	0	0	0	0	1	1	0	0	0	0	0	0	0	0	0	0	0	0	0
A42	0	0	0	0	0	0	0	0	0	0	0	0	0	0	0	1	0	0	0	0
A43	0	0	0	0	0	0	0	0	0	0	0	0	0	1	0	1	0	0	0	0
A44	0	0	0	0	0	0	0	0	0	0	0	0	0	1	0	1	0	0	0	0
A45	0	0	0	0	0	0	0	0	0	0	0	0	0	0	0	0	0	0	0	0
A46	0	0	0	0	0	0	0	0	0	1	0	0	0	0	0	0	0	1	0	0
A47	0	0	0	0	0	0	0	0	0	0	0	0	0	0	0	0	0	0	0	0
A48	0	1	0	0	0	0	0	0	0	0	0	0	0	0	0	0	0	0	0	0
A49	0	0	0	0	0	0	0	0	1	0	0	0	0	1	0	0	0	0	0	0
A50	0	0	0	0	0	0	0	0	0	0	0	0	0	0	0	0	0	0	0	0

ID	A21	A22	A23	A24	A25	A26	A27	A28	A29	A30	A31	A32	A33	A34	A35	A36	A37
A1	0	0	0	1	0	0	0	0	0	0	0	0	0	0	0	0	0
A2	0	0	0	0	0	0	0	0	0	0	0	1	0	1	0	0	0
A3	0	0	0	0	0	0	0	0	0	0	0	0	0	0	0	0	0
A4	0	0	0	1	0	0	0	0	0	1	0	0	0	0	0	0	0
A5	0	0	0	0	0	1	0	0	0	0	0	0	0	0	0	0	0
A6	0	0	0	0	0	1	0	1	0	0	0	0	0	0	0	0	0
A7	0	0	0	0	0	0	0	0	0	0	0	0	0	0	0	0	0
A8	0	0	0	0	1	0	0	0	0	0	0	0	0	0	0	0	0
A9	0	0	0	0	0	0	1	0	0	0	0	0	0	0	0	0	0
A10	0	0	0	0	0	0	0	0	0	0	0	0	0	0	1	0	0
A11	0	0	0	0	0	0	0	0	0	0	0	0	0	0	0	0	1
A12	0	0	0	0	0	0	0	0	0	0	0	0	0	0	0	0	0
A13	0	0	0	0	0	0	0	0	0	0	0	0	0	0	0	0	0
A14	0	0	0	1	0	0	0	0	0	0	0	0	0	1	0	0	0
A15	0	0	0	0	0	0	1	0	0	0	0	0	0	0	0	0	0
A16	0	0	0	0	0	0	0	0	0	0	0	0	0	0	0	0	0
A17	0	0	0	0	0	0	0	0	0	0	0	0	0	0	0	0	0
A18	0	0	0	0	0	0	0	0	0	0	0	0	0	0	0	0	0
A19	1	0	0	0	0	0	0	0	0	0	0	1	0	1	0	0	0
A20	0	0	0	0	0	0	0	0	0	0	0	0	0	0	0	0	0

TABLE B.8 (*Continued*)

ID	A21	A22	A23	A24	A25	A26	A27	A28	A29	A30	A31	A32	A33	A34	A35	A36	A37
A21	0	0	0	0	0	0	0	0	0	0	0	0	0	1	0	0	0
A22	0	0	0	0	0	0	0	0	0	0	0	0	0	0	0	0	0
A23	0	0	0	0	0	0	0	0	0	0	1	0	0	0	0	0	0
A24	0	0	0	0	0	0	0	0	0	0	0	0	0	0	0	0	0
A25	0	0	0	0	0	0	0	0	0	0	0	1	0	1	0	0	0
A26	0	0	0	0	0	0	0	1	0	0	0	0	0	0	0	0	0
A27	0	0	0	0	0	0	0	0	0	0	0	0	0	0	0	0	0
A28	0	0	0	0	0	0	0	0	0	0	0	0	0	0	0	1	0
A29	0	0	0	0	0	0	0	0	0	0	0	0	1	0	0	0	0
A30	0	0	0	0	0	0	0	0	0	0	0	0	0	0	0	0	0
A31	0	0	0	0	1	0	0	0	0	0	0	0	0	0	0	0	0
A32	0	1	0	0	0	0	0	0	0	0	0	0	0	0	0	0	0
A33	0	0	0	0	0	0	0	0	1	0	0	0	0	0	0	0	0
A34	1	0	0	0	0	0	0	0	0	0	0	0	0	0	0	0	0
A35	0	0	0	0	0	0	0	0	0	0	0	0	0	0	0	0	0
A36	0	0	0	0	0	0	0	0	0	0	0	0	0	0	0	0	0
A37	0	0	0	0	0	0	0	0	0	0	0	0	0	0	0	0	0
A38	0	0	0	0	0	0	0	0	0	0	0	0	0	0	0	0	0
A39	0	0	0	0	0	0	0	0	0	0	0	0	0	0	0	0	1
A40	0	0	0	0	0	0	0	0	0	1	0	0	0	0	0	0	0
A41	0	0	0	0	0	0	0	0	0	0	0	0	1	0	0	0	0
A42	0	0	0	0	0	0	1	0	0	0	0	0	0	0	0	0	0
A43	0	0	0	0	0	0	0	0	0	0	0	0	0	0	0	0	0
A44	0	0	0	0	0	0	1	0	0	0	0	0	0	0	0	0	0
A45	0	0	0	0	0	0	0	0	0	0	0	0	0	0	0	0	0
A46	0	0	0	0	0	0	0	0	0	0	0	0	0	0	1	0	0
A47	0	0	0	0	0	0	0	0	0	0	0	0	0	0	0	0	0
A48	1	0	0	0	1	0	0	0	0	0	0	0	0	0	0	0	0
A49	0	0	0	0	0	0	0	0	0	0	0	0	0	0	0	0	0
A50	0	0	0	0	0	0	0	0	0	0	0	0	0	0	0	0	0

(*continued*)

TABLE B.8 (*Continued*)

ID	A38	A39	A40	A41	A42	A43	A44	A45	A46	A47	A48	A49	A50
A1	0	0	0	0	0	0	0	0	0	0	0	0	0
A2	0	0	0	0	0	0	0	0	0	0	1	0	0
A3	0	0	0	0	0	0	0	0	0	0	0	0	0
A4	0	0	0	0	0	1	0	0	0	0	0	0	0
A5	0	1	0	0	0	0	0	0	0	0	0	0	0
A6	0	0	0	0	0	0	0	0	0	0	0	0	0
A7	0	0	0	0	0	0	0	0	0	0	0	0	0
A8	0	0	0	0	0	0	0	0	0	0	0	0	0
A9	0	0	0	0	0	0	1	0	0	0	0	1	0
A10	0	0	0	0	0	0	0	0	0	0	0	0	0
A11	0	0	0	1	0	0	0	0	0	1	0	0	0
A12	0	0	0	0	0	0	0	0	0	0	0	0	0
A13	0	0	0	0	0	0	0	0	0	0	0	0	0
A14	0	0	1	0	0	0	0	0	0	0	0	0	0
A15	0	0	0	0	1	0	0	0	0	0	0	0	0
A16	0	0	0	0	0	1	0	0	0	0	0	0	0
A17	0	0	0	0	0	0	1	0	0	0	0	0	0
A18	0	0	0	0	0	0	0	0	0	0	0	0	0
A19	0	0	0	0	0	0	0	0	0	0	1	0	0
A20	0	0	0	0	0	0	0	0	1	0	0	0	0
A21	0	0	0	0	0	0	0	0	0	0	0	0	0
A22	1	0	0	0	0	0	0	0	0	0	0	0	1
A23	0	0	0	0	0	0	0	0	0	0	0	0	0
A24	0	0	1	0	0	1	0	0	0	0	0	0	0
A25	0	0	0	0	0	0	0	0	0	0	0	0	0
A26	0	0	0	0	0	0	0	0	0	0	0	0	0
A27	0	0	0	0	0	0	0	0	0	0	0	0	0
A28	0	0	0	0	0	0	0	0	0	0	0	0	0
A29	0	0	0	0	0	0	0	0	0	0	0	0	0

TABLE B.8   *(Continued)*

ID	A38	A39	A40	A41	A42	A43	A44	A45	A46	A47	A48	A49	A50
A30	0	0	0	0	0	0	0	0	0	0	0	0	0
A31	0	0	0	0	0	0	0	0	0	0	0	0	0
A32	0	0	0	0	0	0	0	0	0	0	1	0	0
A33	0	0	0	0	0	0	0	0	0	0	0	0	0
A34	0	0	0	0	0	0	0	0	0	0	0	0	0
A35	0	0	0	0	0	0	0	0	0	0	0	0	0
A36	0	0	0	0	0	0	0	0	0	0	0	0	0
A37	0	0	0	0	0	0	0	0	0	0	0	0	0
A38	0	0	0	0	0	0	0	0	0	0	0	0	0
A39	0	0	0	0	0	0	0	1	0	1	0	0	0
A40	0	0	0	0	0	0	0	0	0	0	0	0	0
A41	0	0	0	0	0	0	0	0	0	0	0	0	0
A42	0	0	0	0	0	0	1	0	0	0	0	1	0
A43	0	0	1	0	0	0	0	0	0	0	0	0	0
A44	0	0	0	0	0	0	0	0	0	0	0	1	0
A45	0	0	0	0	0	0	0	0	0	0	0	0	0
A46	0	0	0	0	0	0	0	0	0	0	0	0	0
A47	0	0	0	0	0	0	0	0	0	0	0	0	0
A48	0	0	0	0	0	0	0	0	0	0	0	0	0
A49	0	0	0	0	0	0	1	0	0	0	0	0	0
A50	1	0	0	0	0	0	0	0	0	0	0	0	0

TABLE B.9 Lab Exercise 5 Agent Attributes

Agent No.	Project	Agent No.	Project
Agent_1	2	Agent_26	6
Agent_2	2	Agent_27	2
Agent_3	5	Agent_28	7
Agent_4	2	Agent_29	4
Agent_5	2	Agent_30	2
Agent_6	6	Agent_31	1
Agent_7	1	Agent_32	2
Agent_8	5	Agent_33	4
Agent_9	2	Agent_34	3
Agent_10	6	Agent_35	5
Agent_11	3	Agent_36	4
Agent_12	2	Agent_37	4
Agent_13	7	Agent_38	6
Agent_14	3	Agent_39	3
Agent_15	4	Agent_40	7
Agent_16	3	Agent_41	1
Agent_17	3	Agent_42	3
Agent_18	6	Agent_43	4
Agent_19	4	Agent_44	3
Agent_20	7	Agent_45	3
Agent_21	3	Agent_46	7
Agent_22	5	Agent_47	5
Agent_23	1	Agent_48	4
Agent_24	4	Agent_49	2
Agent_25	2	Agent_50	5

TABLE B.10 Lab Exercise 7–Agent × Location Network

ID	L1	L2	L3	L4	L5	L6	L7	L8	L9	L10
A1	1	1	1	1	0	0	0	0	0	0
A2	0	0	0	1	1	0	0	0	0	1
A3	1	0	0	0	1	0	0	1	0	0
A4	0	0	1	1	0	1	0	0	0	0
A5	1	0	0	0	0	0	1	0	0	0
A6	0	0	0	1	0	0	1	0	0	0
A7	0	1	0	0	0	0	0	1	0	0
A8	0	0	0	0	1	0	0	1	0	0
A9	0	1	0	1	0	0	0	1	0	1
A10	0	0	0	0	0	0	0	1	1	0
A11	0	0	1	1	0	0	1	0	0	0
A12	0	0	1	0	1	0	0	0	0	0
A13	1	1	1	1	0	0	0	0	1	0
A14	0	0	0	1	0	0	0	0	0	0
A15	0	0	1	0	0	1	0	1	0	0

TABLE B.11   Lab Exercise 7–Agent × Resource Network

ID	R1	R2	R3	R4	R5	R6	R7	R8	R9	R10
A1	1	1	1	0	0	0	0	0	0	0
A2	0	0	0	0	0	1	1	0	1	0
A3	0	0	0	0	0	0	1	0	0	0
A4	0	0	1	0	0	1	0	0	0	0
A5	1	0	0	0	0	0	0	1	0	0
A6	0	0	0	0	0	0	0	1	0	0
A7	0	1	0	0	0	0	0	1	0	0
A8	0	0	0	0	0	0	0	1	0	0
A9	0	1	0	0	0	0	0	0	1	1
A10	0	0	0	0	0	0	0	1	0	1
A11	0	0	1	0	0	0	0	0	0	0
A12	0	0	0	0	0	0	1	0	0	0
A13	0	0	1	1	1	0	0	0	0	0
A14	0	0	0	0	0	0	0	0	0	0
A15	0	0	0	0	0	0	0	1	0	0

TABLE B.12   Lab Exercise 7 — Resource × Agent Network

ID	A1	A2	A3	A4	A5	A6	A7	A8	A9	A10	A11	A12	A13	A14	A15
R1	0	0	0	0	1	0	0	0	0	0	0	0	0	0	0
R2	0	0	0	0	0	0	1	0	1	0	0	0	0	0	0
R3	0	0	0	1	0	0	0	0	0	0	0	1	0	0	0
R4	0	0	1	0	0	0	0	0	1	0	0	0	0	0	0
R5	0	1	0	0	0	0	0	0	0	0	0	0	0	0	0
R6	0	1	0	0	0	0	0	0	0	0	0	0	0	0	0
R7	0	1	0	0	0	0	0	0	0	0	0	0	0	0	0
R8	0	1	0	0	0	0	0	0	0	0	1	0	0	1	0
R9	0	0	1	0	0	0	0	0	0	0	0	0	0	0	0
R10	0	1	0	0	0	0	0	0	0	0	0	0	0	0	0

TABLE B.13   Lab Exercise 7 — Location × Location Network

ID	L1	L2	L3	L4	L5	L6	L7	L8	L9	L10
L1	0	0	0	0	1	0	0	0	0	0
L2	0	0	0	0	0	0	0	1	0	0
L3	0	0	0	1	0	0	0	0	0	0
L4	0	0	1	0	0	1	0	0	0	0
L5	1	0	0	0	0	0	0	0	0	0
L6	0	0	1	1	0	0	0	0	0	0
L7	0	0	0	0	0	0	0	0	1	0
L8	0	1	0	0	0	0	0	0	0	1
L9	0	0	0	0	0	0	1	0	0	0
L10	0	0	0	0	0	0	0	1	0	0

TABLE B.14   Lab Exercise 7 — Resource × Resource Network

ID	R1	R2	R3	R4	R5	R6	R7	R8	R9	R10
R1	0	1	1	0	0	0	0	0	0	0
R2	0	0	0	0	0	0	0	0	0	0
R3	0	0	0	1	1	0	0	0	0	0
R4	0	0	0	0	0	0	0	0	0	0
R5	0	0	0	0	0	0	0	0	0	0
R6	0	0	0	0	0	0	0	0	0	1
R7	0	0	0	0	0	0	0	0	0	1
R8	0	0	0	0	0	0	0	0	0	1
R9	0	0	0	0	0	0	0	0	0	1
R10	0	0	0	0	0	0	0	0	0	0

TABLE B.15   Lab Exercise 9 — Formal Social Network: Agent × Agent

ID	A1	A2	A3	A4	A5	A6	A7	A8	A9	A10	A11	A12	A13	A14	A15	A16	A17	A18	A19	A20
A1		1	1	1																
A2	1		1	1									1							
A3	1	1													1				1	1
A4	1						1													
A5							1													
A6							1													
A7								1												
A8								1												
A9								1												
A10																				
A11													1							
A12													1							
A13														1						
A14														1						
A15															1					
A16															1					
A17																1				
A18																1				
A19																				
A20																				

TABLE B.16   Lab Exercise 9 — Informal Social Network: Agent × Agent

	A1	A2	A3	A4	A5	A6	A7	A8	A9	A10	A11	A12	A13	A14	A15	A16	A17	A18	A19	A20
A1																				
A2	1		1			1														
A3	1	1		1		1														
A4		1	1																	
A5							1													
A6							1		1		1									
A7							1				1									
A8																				
A9																				
A10			1		1				1											
A11					1															
A12				1																
A13															1		1			1
A14															1		1			
A15															1	1				
A16															1	1				
A17															1					
A18															1					
A19															1					
A20															1					

TABLE B.17   Lab Exercise 9 — Agent × Knowledge Network Indicating Which Agent Holds What Knowledge

	K1	K2	K3	K4	K5	K6	K7	K8	K9	K10	K11	K12	K13	K14	K15	K16
A1		1	1					1								
A2					1		1	1								
A3										1	1					
A4														1		
A5												1				
A6												1	1			
A7																
A8														1		
A9													1			
A10						1										
A11												1				
A12												1				
A13						1										
A14			1													
A15																
A16																
A17															1	
A18	1															
A19		1														
A20		1														

TABLE B.18   Lab Exercise 9 — Agent × Task Network Indicating the Task Each Agent has been Assigned

	T1	T2	T3	T4	T5	T6	T7	T8	T9	T10	T11	T12	T13	T14	T15	T16	T17	T18	T19	T20
A1												1			1					1
A2			1		1									1						
A3																		1		
A4	1	1	1	1																
A5		1																		
A6	1	1																		
A7	1																			
A8	1																			
A9					1								1							
A10						1								1						
A11		1																		
A12			1																	
A13				1																
A14																		1		
A15								1												
A16							1													
A17								1												
A18											1									
A19											1									
A20											1									

TABLE B.19   Lab Exercise 9 — Knowledge × Task Network Showing What Knowledge is Required to Carry Out Which Tasks

	T1	T2	T3	T4	T5	T6	T7	T8	T9	T10	T11	T12	T13	T14	T15	T16	T17	T18	T19	T20
K1																				
K2																				1
K3												1			1					
K4						1												1		
K5				1																
K6				1																
K7				1	1	1														
K8				1		1														
K9		1	1								1									1
K10									1											
K11									1											
K12		1	1								1									
K13	1	1				1					1									
K14	1	1		1																
K15								1								1	1	1		
K16																				1

FIVE POINTS OF A GRAPH

There are 34 permutations of a five node network.

Graph	C_D	C_D (n_i)	C_D^* (n_i)	C_B	C_B (n_i)	C_B^* (n_i)	C_C	C_C^{-1} (n_i)	C_C^* (n_i)
1.	0	0	0	0	0	0	*	*	*
		0	0		0	0		*	*
		0	0		0	0		*	*
		0	0		0	0		*	*
		0	0		0	0		*	*
Null									
2.	0.25	1	0.25	0	0	0	*	*	*
		1	0.25		0	0		*	*
		0	0		0	0		*	*
		0	0		0	0		*	*
		0	0		0	0		*	*
3.	0.50	2	0.50	0.16	1	0.16	*	*	*
		1	0.25		0	0		*	*
		1	0.25		0	0		*	*
		0	0		0	0		*	*
		0	0		0	0		*	*

(continued)

Social Network Analysis: with Applications, First Edition.
Ian A. McCulloh, Helen L. Armstrong, and Anthony N. Johnson.
© 2013 John Wiley & Sons, Inc. Published 2013 by John Wiley & Sons, Inc.

Graph	C_D	C_D (n_i)	C_D^* (n_i)	C_B	C_B (n_i)	C_B^* (n_i)	C_C	C_C^{-1} (n_i)	C_C^* (n_i)
4.	0.08	1	0.25	0	0	0	*	*	*
		1	0.25		0	0		*	*
		1	0.25		0	0		*	*
		1	0.25		0	0		*	*
		0	0		0	0		*	*
5.	0.75	3	0.75	0.50	3	0.50	*	*	*
		1	0.25		0	0		*	*
		1	0.25		0	0		*	*
		1	0.25		0	0_		*	*
		0	0		0	0		*	*
6.	0.33	2	0.50	0.17	1	0.17	*	*	*
		1	0.25		0	0		*	*
		1	0.25		0	0		*	*
		1	0.25		0	0		*	*
		1	0.25		0	0		*	*
7.	0.33	2	0.50	0	0	0	*	*	*
		2	0.50		0	0		*	*
		2	0.50		0	0		*	*
		0	0		0	0		*	*
		0	0		0	0		*	*
8.	0.33	2	0.50	0.25	2	0.33	*	*	*
		2	0.50		2	0.33		*	*
		1	0.25		0	0		*	*
		1	0.25		0	0		*	*
		0	0		0	0		*	*

Graph	C_D	C_D (n_i)	C_D^* (n_i)	C_B	C_B (n_i)	C_B^* (n_i)	C_C	C_C^{-1} (n_i)	C_C^* (n_i)
9.	1	4	1	1	6	1	1	4	1
		1	0.25		0	0		7	.57
		1	0.25		0	0		7	.57
		1	0.25		0	0		7	.57
		1	0.25		0	0		7	.57
Star									
10.	0.58	3	0.75	0.71	5	0.83	0.63	5	0.80
		2	0.50		2	0.50		6	0.67
		1	0.25		0	0		8	0.50
		1	0.25		0	0		8	0.50
		1	0.25		0	0		9	0.44
Fork									
11.	0.17	2	0.50	0.41	4	0.67	0.43	6	0.67
		2	0.50		3	0.50		7	0.57
		2	0.50		3	0.50		7	0.57
		1	0.25		0	0		10	0.40
		1	0.25		0	0		10	0.40
Chain									
12.	0.58	3	0.75	0.33	2	0.33	*	*	*
		2	0.50		0	0		*	*
		2	0.50		0	0		*	*
		1	0.25		0	0		*	*
		0	0		0	0		*	*
13.	0.17	2	0.50	0	0	0	*	*	*
		2	0.50		0	0		*	*
		2	0.50		0	0		*	*
		1	0.25		0	0		*	*
		1	0.25		0	0		*	*

(continued)

Graph	C_D	C_D (n_i)	C_D^* (n_i)	C_B	C_B (n_i)	C_B^* (n_i)	C_C	C_C^{-1} (n_i)	C_C^* (n_i)
14.	0.17	2	0.50	0.20	0.50	0.08	*	*	*
		2	0.50		0.50	0.08		*	*
		2	0.50		0.50	0.08		*	*
		2	0.50		0.50	0.08		*	*
		0	0		0	0		*	*
15.	0.83	4	1	0.83	5	0.83	0.89	4	1
		2	0.50		0	0		6	0.67
		2	0.50		0	0		6	0.67
		1	0.25		0	0		7	0.57
		1	0.25		0	0		7	0.57
16.	0.42	3	0.75	0.38	3	0.50	0.43	5	0.80
		3	0.75		3	0.50		5	0.80
		2	0.50		0	0		6	0.67
		1	0.25		0	0		8	0.50
		1	0.25		0	0		8	0.50
17.	0.42	3	0.75	0.56	4	0.67	0.55	5	0.80
		2	0.75		3	0		6	0.67
		2	0.50		0	0		7	0.57
		2	0.50		0	0		7	0.57
		1	0.25		0	0		9	0.44
18.	0.42	3	0.75	0.48	3.50	0.58	0.46	5	0.80
		2	0.75		1	0.17		6	0.67
		2	0.50		1	0.17		6	0.67
		2	0.50		0.50	0.08		7	0.57
		1	0.25		0	0		8	0.50

Graph	C_D	C_D (n_i)	C_D^* (n_i)	C_B	C_B (n_i)	C_B^* (n_i)	C_C	C_C^{-1} (n_i)	C_C^* (n_i)
19.	0.42	3	0.75	0.06	0.50	0.08	*	*	*
		3	0.75		0.50	0.08		*	*
		2	0.50		0	0		*	*
		2	0.50		0	0		*	*
		0	0		0	0		*	*
20.	0	2	0.50	0	1	0.17	0	6	0.67
		2	0.50		1	0.17		6	0.67
		2	0.50		1	0.17		6	0.67
		2	0.50		1	0.17		6	0.67
		2	0.50		1	0.17		6	0.67
Circle **21.**	0.67	4	1	0.67	4	0.67	0.77	4	1
		2	0.50		0	0		6	0.67
		2	0.50		0	0		6	0.67
		2	0.50		0	0		6	0.67
		2	0.50		0	0		6	0.67
22.	0	0	0	0	0	0	*	*	*
		0	0		0	0		*	*
		0	0		0	0		*	*
		0	0		0	0		*	*
		0	0		0	0		*	*
23.	0	0	0	0	0	0	*	*	*
		0	0		0	0		*	*
		0	0		0	0		*	*
		0	0		0	0		*	*
		0	0		0	0		*	*

(continued)

Graph	C_D	C_D (n_i)	C_D^* (n_i)	C_B	C_B (n_i)	C_B^* (n_i)	C_C	C_C^{-1} (n_i)	C_C^* (n_i)
24.	0	0	0	0	0	0	*	*	*
		0	0		0	0		*	*
		0	0		0	0		*	*
		0	0		0	0		*	*
		0	0		0	0		*	*
25.	0	0	0	0	0	0	*	*	*
		0	0		0	0		*	*
		0	0		0	0		*	*
		0	0		0	0		*	*
		0	0		0	0		*	*
26.	0	0	0	0	0	0	*	*	*
		0	0		0	0		*	*
		0	0		0	0		*	*
		0	0		0	0		*	*
		0	0		0	0		*	*
27.	0	0	0	0	0	0	*	*	*
		0	0		0	0		*	*
		0	0		0	0		*	*
		0	0		0	0		*	*
		0	0		0	0		*	*
28.	0	0	0	0	0	0	*	*	*
		0	0		0	0		*	*
		0	0		0	0		*	*
		0	0		0	0		*	*
		0	0		0	0		*	*

Graph	C_D	C_D (n_i)	C_D^* (n_i)	C_B	C_B (n_i)	C_B^* (n_i)	C_C	C_C^{-1} (n_i)	C_C^* (n_i)
29.	0	0	0	0	0	0	*	*	*
		0	0		0	0		*	*
		0	0		0	0		*	*
		0	0		0	0		*	*
		0	0		0	0		*	*
30.	0	0	0	0	0	0	*	*	*
		0	0		0	0		*	*
		0	0		0	0		*	*
		0	0		0	0		*	*
		0	0		0	0		*	*
31.	0	0	0	0	0	0	*	*	*
		0	0		0	0		*	*
		0	0		0	0		*	*
		0	0		0	0		*	*
		0	0		0	0		*	*
32.	0	0	0	0	0	0	*	*	*
		0	0		0	0		*	*
		0	0		0	0		*	*
		0	0		0	0		*	*
		0	0		0	0		*	*
33.	0	0	0	0	0	0	*	*	*
		0	0		0	0		*	*
		0	0		0	0		*	*
		0	0		0	0		*	*
		0	0		0	0		*	*

(*continued*)

Graph	C_D	C_D (n_i)	C_D^* (n_i)	C_B	C_B (n_i)	C_B^* (n_i)	C_C	C_C^{-1} (n_i)	C_C^* (n_i)
34.	0	0	0	0	0	0	*	*	*
		0	0		0	0		*	*
		0	0		0	0		*	*
		0	0		0	0		*	*
		0	0		0	0		*	*

Complete

INDEX

Social Network Analysis: with Applications, First Edition.
Ian A. McCulloh, Helen L. Armstrong, and Anthony N. Johnson.
© 2013 John Wiley & Sons, Inc. Published 2013 by John Wiley & Sons, Inc.